Smart Microgrids

Smart Microgrids

Edited by
Sasi K. Kottayil

CRC Press
Taylor & Francis Group
Boca Raton London New York

CRC Press is an imprint of the
Taylor & Francis Group, an **informa** business

MATLAB® is a trademark of The MathWorks, Inc. and is used with permission. The MathWorks does not warrant the accuracy of the text or exercises in this book. This book's use or discussion of MATLAB® software or related products does not constitute endorsement or sponsorship by The MathWorks of a particular pedagogical approach or particular use of the MATLAB® software.

First edition published 2020
by CRC Press
6000 Broken Sound Parkway NW, Suite 300, Boca Raton, FL 33487-2742

and by CRC Press
2 Park Square, Milton Park, Abingdon, Oxon, OX14 4RN

ISBN: 978-0-367-34362-0 (hbk)
ISBN: 978-0-367-53446-2 (pbk)
ISBN: 978-0-429-32527-4 (ebk)

Typeset in Times
by Lumina Datamatics Limited

Dedicated to

———————

Our Great Teachers,
Professor Sujay Basu (late) &
Professor A. T. Devarajan

Contents

Foreword

When Prof. Sasi K. Kottayil requested me to write a Foreword for this book on *Smart Microgrids*, I remembered the many discussions we had on the topic of renewable energy over the last 12 years. We asked a number of questions with regard to situation in rural India where there are several areas not "fully electrified" or where availability of electricity from the grid is intermittent. These areas are moderately rich in energy sources, in the form of solar energy, biomass, and wind. If only there were sufficient investment available and cost could be reasonable, one could think of energy generation from these various sources (polygeneration) forming into grid-connectable microgrids and nesting them together from village to village. Of course, there was the question of appropriate methods and technology, which we felt is only a matter for development and not a real obstacle.

The first structured collaboration between KTH and Amrita University research groups on microgrids started in 2010 with a three-year collaborative research project funding jointly by the department of science and technology in India and Vinnova, Sweden. The topic was "Energy Management on Smart Grid Using Embedded Systems." That project resulted in several joint workshops and visit of researchers in both directions. Another joint effort in microgrids between KTH and team of Prof. Sasi K. Kottayil was the feasibility study of renewable energy generation at Andaman and Nicobar Islands during 2015–2016 with funding from Swedish Energy Agency, in cooperation with A&N islands renewable energy authority. During that time, a 30-kVA grid-connectable demonstration microgrid involving wind energy, pumped hydro, and biomass was proposed in Port Blair. Even though that project was not implemented, the project was a motivation for investigating many technologies related to microgrid.

This book on *Smart Microgrids* by Prof. Sasi K. Kottayil and his team covers most aspects of the planning, design, control, and operation of grid-connectable microgrid. This book will be a valuable resource, both for engineering students, and practicing engineers interested in microgrids.

Rajeev Thottappillil, PhD
Professor in Electric Power Engineering and Design
Director of SweGRIDS (www.swegrids.se)
School of Electrical Engineering and Computer Science
KTH Royal Institute of Technology, Stockholm
Sweden

Preface

When the twenty-first century leaves back its teen-ages, the scientists and engineers all over the world are busy engaged in shaping and consolidating the fourth industrial revolution. Yet, everything is changing and the major cause behind all the changes, the commercial energy systems, is the one that has metamorphosed the most at this point of time. The traditional power system with centralized generation and synchronous distribution is going through the revolutionary change, may be to evolve finally (?) as asynchronously linked and geographically integrated macro grid – a conglomeration of large and small grids. Though the development happens in power system, the devices involved are partly of power electronics and the rest is of information and communication technology. The integration uses advanced soft computing techniques, but the most binding factor is power balancing and the most powerful tool is control theory. The amazing growth of engineering knowledge has made it happen – microgrid has come of age!

This book, *Smart Microgrids*, is a joint effort by four authors to help power engineering students, both graduate and undergraduate, to interlay the knowledge blocks with an aspiration to build the sustainable grid of tomorrow.

The readers can find the introduction to the topics in Chapter 1. Chapter 2 takes you through the intricacies of microgrid design and operation. The basics of system integration and the concepts of power converter controls shall help understanding microgrids scrupulously. Communication network on the smart power systems is the topic of discussion in Chapter 3. It describes the protocols and architecture to help identify what is used where. ICT applications on power grid are introduced in this chapter. A recent development of ICT based demand side management, namely demand dispatch, is presented in Chapter 4. It is the dissemination of a research on demand dispatch framework which includes its method of validation too. Chapter 5 presents yet another ICT application – dynamic energy management system. The author has shared the experience of learning by making that technology.

The conceptual understanding of microgrid infrastructure development and its operation in the second chapter and the potential support on tools and information of communication superstructure in the third chapter, we hope, will help power engineering students to amply comprehend this smart technology, and, the comprehensive demonstrations presented in the fourth and fifth chapters will motive them to venture on more meaningful projects and research. The information shared and the concepts presented are expected to be helpful to practicing engineers too.

This book is the upshot of decade long learning by the four authors on individual portfolios in a common space – the research project, *Energy Management on Smart Grid using Embedded Systems*, funded by DST (India) and VINNOVA (Sweden) during 2011–2014 preluded this learning. Graduate students' projects supervised by the authors and sponsored research involved have immensely contributed to the confidence level of authorship showcased here. We place on record our sincere gratitude to all those who supported this endeavor in work and spirit.

The final word of gratitude is to Taylor & Francis Group, more for the encourage-
ment than for the facilitation.

Sasi K. Kottayil

A. Vijayakumari

P. Sivraj

S. Nithin

D. Prasanna Vadana

MATLAB® is a registered trademark of The MathWorks, Inc. For product informa-
tion, please contact:
  The MathWorks, Inc.
  3 Apple Hill Drive
  Natick, MA 01760-2098 USA
  Tel: 508 647 7000
  Fax: 508-647-7001
  E-mail: info@mathworks.com
  Web: www.mathworks.com

Editor

Sasi K. Kottayil, born in 1957 in India, earned both bachelor's and master's degrees in the field of electrical engineering and a doctoral degree on *utility scale wind electricity generation*. He has been in the teaching profession since 1984. Renewable energy and smart microgrid are his areas of interest. He is currently a professor in the department of electrical engineering at Amrita Vishwa Vidyapeetham, Coimbatore, India.

Contributors

Sasi K. Kottayil
Department of Electrical & Electronics
 Engineering
Amrita School of Engineering
Amrita Vishwa Vidyapeetham
 University
Coimbatore, India

S. Nithin
Department of Electrical & Electronics
 Engineering
Amrita School of Engineering
Amrita Vishwa Vidyapeetham
 University
Coimbatore, India

D. Prasanna Vadana
Department of Electrical & Electronics
 Engineering
Amrita School of Engineering
Amrita Vishwa Vidyapeetham
 University
Coimbatore, India

P. Sivraj
Department of Electrical & Electronics
 Engineering
Amrita School of Engineering
Amrita Vishwa Vidyapeetham
 University
Coimbatore, India

A. Vijayakumari
Department of Electrical & Electronics
 Engineering
Amrita School of Engineering
Amrita Vishwa Vidyapeetham
 University
Coimbatore, India

List of Abbreviations

3G	Third generation
3GPP	Third Generation Partnership Project
6LoWPAN	IPv6 over low power wireless personal area networks
AC	alternating current
ADS	aggregator dispatch share
ADSA	aggregator dispatch share allocation
AES	advance encryption standard
AFC	automatic frequency control
AI	artificial intelligence
AMC	adaptive modulation and coding
AMI	advanced metering infrastructure
AMR	automatic meter reading
ANN	artificial neural networks
ANSI	American National Standards Institute
AODV	ad-hoc on-demand distance vector
ARQ	automatic retransmission requests
ASEAN	Association for South East Asian Nations
BAN	building area networks
BES	battery energy storage
BNEP	Bluetooth network encapsulation protocol
BP	back propagation
BPL	broadband over power line
BPSK	binary phase shift keying
BSS	basic service set
C	capacitor
CA	collision avoidance
CAN	controller area network
CEMS	centralized energy management system
CDMA	code division multiple access
CHP	combined heat and power
CIGRÉ	Conseil International des Grands Réseaux Électriques
COAP	constrained application protocol
COMSEM	companion specification for energy metering
COP	conference of the parties
CR	cognitive radio
CRC	cyclic redundancy check
CSL	coordinated sampled listening
CSMA	carrier sense multiple access
CT-IAP	communication technology integrated architectural perspectives

CU	control units
DA	distributed automation
DAG	direct acyclic graph
De-CEMS	decentralized CEMS
DC	direct current or data concentrators (depending on context)
DD	demand dispatch
DDA	demand dispatch aggregator
DDP	demand dispatch provider
DEM	dynamic energy management
DER	distributed energy resources
DFIG	doubly fed induction generator
DG	distributed generation
DGM	distribution grid management
DKE	German Commission for Electrical and Electronic and Information Technologies of DIN and VDE
DL	dispatchable load
DLC	direct load control
DLMS	device language message specification
DMS	distribution management systems
DNO3	distributed network protocol 3
DPS	differential phase-shift keying
DR	demand response
DSF	double synchronous frame
DQPSK	differential quadrature phase-shift keying
DSM	demand side management
DSSS	direct sequence spread spectrum
DTLS	datagram transport layered security
EAP	extensible authentication protocol
EDFA	erbium-doped fiber amplifier
EDGE	enhanced data rates for GSM evolution
EMD	empirical mode decomposition
EMS	energy management system
ESCON	enterprise system connection
ESI	energy service interface
ESS	energy storage systems or extended service set (depending on context)
ETSI	European Telecommunications Standards Institute
EVSE	electric vehicle supply equipment
EWT	empirical wavelet transforms
FAN	field area networks
FDDI	fiber distributed data interface
FDM	frequency division multiplexing
FFD	full function device
FOCV	fractional open circuit voltage
FPGA	field programmable gate array
FSCC	fractional short circuit current
FSK	frequency shift keying

FSSS	frequency hopping spread spectrum
G2V	power movement from grid to electric vehicles
GFSK	Gaussian frequency shift keying
GHG	greenhouse gas
GPRS	general packet radio service
GPS	global positioning system
GSM	global system for mobile
GUI	graphical user interface
HAN	home area networks
HART	highway addressable remote transducer protocol
HIL	hardware in the loop
HMI	human machine interface
HSDPA	high speed downlink packet access
HSPA	high speed packet access
HSPA+	evolved high speed packet access
HSUPA	high speed uplink packet access
Hydel	hydroelectric
IAN	industry area network
IAP	integrated architectural perspectives
IC	incremental conductance
ICT	information and communication technology
ID	islanding detection
IEC	International Electrotechnical Commission
IED	intelligent electronic device
IEEE	Institute of Electrical and Electronics Engineers
IETF	internet engineering task force
INDC	Intended Nationally Determined Contributions (as delineated at U.N. climate talks)
IP	internet protocol
IPCC	Intergovernmental Panel on Climate Change
IPP	independent power producers
ISM	industrial scientific and medicinal
ISO	independent system operators or International Standards Organization (depending on context)
ISP	inter-system protocol
IT-IAP	information technology integrated architectural perspectives
KVL	Kirchhoff's voltage law
L	inductor
L2CAP	logical link control and adaptation protocol
LAN	local area network
LC	local controller
LCU	load control unit
LD	local demand
LED	light-emitting diodes
LF	loop filter
LG	local generation

LLC	logical link control
LLNs	low power and lossy networks
L/HVRT	low/high voltage ride-through
LM	load management
LMP	link manager protocol
LOS	line of sight
LTE	long term evolution
LTE-A	LTE-advanced
LUT	look-up tables
MAC	Medium access control
MAN	metropolitan area network
MCS	multi-component signals
MHPP	micro hydro power plant
MIMO	multiple-input, multiple-output
MMF	multi-mode fiber
MOSFET	metal oxide semiconductor field effect transistors
MPC	model predictive controllers
MPP	maximum power point
MPPT	maximum power point tracking
MQTT	message queue telemetry transport
NAN	neighborhood area networks
NaS	Sodium Sulphur (battery)
NDZ	non-detection zones
NIC	network interface card
NIST	National Institute of Standards and Technology
NLOS	non-line of sight
NN	neural network
ND	Newton-Raphson iterative load flow analysis
NRZ	non-return to zero
OAA	one-against-all
OAO	one-against-one
OBIS	object identification system
OCPP	open charge point protocol
OFDM	orthogonal frequency division multiplexing
OFP/UFP	over/under frequency protection
OSI	open systems interconnection
OT	optimum torque
OVP/UVP	over/under voltage protection
P	active power
P&O	perturb and observe
PAN	personal area network
PCC	point of common coupling
PD	phase detect
PDC	phasor data concentrators
PDU	protocol data units
PE	power electronics

PES	power and energy society
PHS	pumped hydro storage
PI	proportional integral
PID	proportional integral derivative
PLC	programmable logic controllers
PLCC	power line carrier communications
PLL	phase locked loop
PMSG	permanent magnet synchronous generator
PMU	phasor measurement unit
POF	plastic/polymer optical fiber
PPP	point to point protocol
PR	proportional resonant
PS-IAP	power systems integrated architectural perspectives
PSO	particle swarm optimization
PV	photovoltaic
PWM	pulse width modulation
Q	reactive power
QoS	quality of service
R	resistor
RBF	radial base function
RE	renewable energy
RFCOMM	radio frequency communication
RFD	reduced function device
ROCOF	rate of change of frequency
RSSI	received signal strength indicator
RTDS	real-time digital simulators
RTO	regional transmission system operators
RTU	remote terminal units
RTDCU	real-time data collection units
SAS	substation automation system
SBCON	serial byte connection
SCADA	supervisory control and data acquisition
SCIG	squirrel cage induction generators
SDH	synchronous data hierarchy
SDP	self-discovery protocol
SFR	stationary reference frame
SKKE	symmetric-key key establishment
SMC	sliding mode controller
SMF	single-mode fiber
SMG	smart microgrids
SMGS	smart microgrid simulator
SoC	state of change
SOFDMA	scalable orthogonal frequency-division multiple access
SOGI	second order generalized integrator
SONET	synchronous optical network
SPP	solar power plant

SRF	synchronous reference frame
SSID	service set identifier
SSP	secure simple pairing
SSSC	*static synchronous series compensator*
STATCOM	*static compensator*
SVC	static VAR compensator
SVM	support vector machines
T&D	transmission and distribution
TCP	transmission control protocol
TCP/IP	transmission control protocol and internet protocol
TCR	thyristor controlled reactor
TDD	total demand distortion
TDMA	time division multiple access
THD	total harmonic distortion
TKIP	temporal key integrity protocol
TLS	transport layer security
TSC	thyristor switched capacitors
TSCH	time slotted channel hopping
UDP	user datagram protocol
UMTS	universal mobile telecommunications system
UNFCCC	United Nations Framework Convention on Climate Change
UPFC	unified power flow controllers
UPQC	unified power quality controller
USIM	universal subscriber identity module
UUID	universally unique identifier
V2G	power movement from electric vehicles to the grid
VAR	volt-ampere reactive (unit of measurement of reactive power)
VCO	voltage control oscillator
VMD	variational mode decomposition
VPN	virtual private network
VSC	voltage source converter
VSI	voltage source inverters
WACS	wide area control systems
WAMPAC	wide area monitoring protection and control
WAMS	wide area measurement systems
WAN	wide area networks
WAPS	wide area protection systems
WASA	wide area situational awareness
WBG	wide band gap
WCDMA	wideband code division multiple access
WDM	wavelength dimension multiplexing
WEP	wired equivalent privacy
WiMAX	worldwide interoperability for microwave access
WPA	Wi-Fi protected access

WPP	wind power plant
WT	wind turbine
WTG	wind turbine generator
ZC	ZigBee coordinator
ZDO	ZigBee device objects
ZED	ZigBee end device
ZR	ZigBee router

1 Introduction

D. Prasanna Vadana, Sasi K. Kottayil,
A. Vijayakumari, P. Sivraj, and S. Nithin

CONTENTS

1.1 ENERGY AND ENVIRONMENT

With the advent of electricity, the world was able to be illuminated on-demand. Thereby, the complex electrical network – now known as the *Grid* – began. This spark enlightened human minds to invent new technologies in various fields which are ever enjoyable and always innovative. Access to electricity has come to be on the list of mandatory needs of a common man in the modern world. This laid the foundation for massive industrial growth, with millions of people benefiting from a safe and reliable supply of electricity from the legacy grid. Yet, the demand for electricity has increased many times over, since every new technology seems to depend on electricity for its operation.

The power grid's intertwined technical, profitable and regulatory substratum was established more than a century ago and has undergone only minimal interference in the decades since. At present, this industry is facing tremendous change. For about a century, affordable electrification was made available to customers by setting up huge power plants of large megawatt generation capability whose power output had to be transported over long transmission lines. Recent technological advancements, coupled with increasing social awareness, are complicating this simple model. People are realizing that we have entered an Anthropocene period of life on Earth, characterized by human impact on global climate change and ecology. And, new technology is making us aware of the actual extent of the devastating impacts that humans are having on the planet.

In recent years, there has been an awful increase in greenhouse gas (GHG) emissions all around the world, for which every individual's recklessness has to be held responsible. This effect ruins the elixir of all living species on this earth – the very breath of life. The rise in CO_2 level over the centuries shows a colossal increase especially from the twentieth to the twenty-first century – a very short time period indeed.

While about 45% of global CO_2 emissions are due to fossil fuel combustion, another 30% is emitted from coal-fired power plants. Consequently, generation of electricity has lead to a solemn commotion in Mother Nature's schedule in the past several years and this is augmented by the fact that many people are still not enjoying the benefits of electricity. Therefore, the increasing demand for electricity will not cease.

Drastic reduction in GHG emissions is possible to a large extent particularly in the electricity sector, when power is generated using non-depletable and non-polluting renewable energy (RE) sources – such as wind, solar, hydro, biomass, etc., as a replacement for the depletable fossil fuels. The eventual rationalization for the use of alternate energy focuses on its alleviation of global warming. Yet, the virtue of this replacement must be coordinated on a global scale, as the penetration of such RE sources into the legacy grid requires decades of expensive dedication.

Nations in the world declared their Intended Nationally Determined Contributions (INDC) during COP 21 of UNFCCC in Paris in December 2015. Later the IPCC declared (at COP 24 in Poland in December 2018) that NDCs as per the Paris Agreement, even if achieved, were not enough to avert the impending climate disaster. Considering the need to limit the temperature rise to 1.5 degree Celsius, IPCC suggested increasing the share of RE-based electricity to 60% by 2030 and to 77% by 2050. It means that small power generation distributed across the nations will have to penetrate the public power grid in a large way, demanding new concepts of distributed control. Utility scale distributed generation (DG) from renewable resources has become mandatory both globally and nationally.

Uncontrolled power generation from intermittent sources like solar and wind can upset the conventional control concepts of traditional power grids, if their penetration levels are high. Not only do these penetration levels need to be high to meet the environmental targets and to maintain renewable portfolio standards, small-scale DG schemes need to be grid-connected too. Penetration of RE sources into the conventional grid poses various challenges in power system management owing to the grid's inherent nature. These challenges can be addressed by integrating smart and intelligent devices throughout the grid having real time monitoring, control and bidirectional communication capabilities. This presents a sophisticated grid capable of expressing its status at any time instantaneously, ultimately transforming "the grid" into a *"smart grid."*

1.2 MICROGRID

One form of RE utilization allows small power production on the distribution network resulting in the formation of microgrids capable of operating in grid-connected and islanded modes. Microgrids are decentralized and self-sustainable electric distribution networks of the future, capable of meeting the local electricity demand in a region utilizing local RE sources, thereby operating as a single controllable system. Each well-designed microgrid, being a deregulated power system, can avert the installation of large-scale power plants. It is relatively straightforward to build a microgrid confined to a region rather than elevating the entire grid to the level of a smart grid, as the latter is a centralized complex network spread globally which may be resistant to change. However, penetration of microgrids onto the main grid reinforces the evolution of the traditional grid into a smarter one, eventually.

Microgrids are constructed with diverse structures including DC, AC and hybrid schemes with combinations of power electronic circuits and networked intelligent controllers so as to accommodate different services for RE plants, energy storage systems (ESS), and consumer integration into a common network. These are designed to operate in various modes; for example, either *grid-connected mode* or *autonomous mode*. But of late, *grid-support mode* has been identified as an ancillary service extended by microgrids to the utility grid especially during fault conditions. The operational control of all these operating modes is such that seamless sliding from any mode to any other is envisaged.

The distinctive attribute of microgrids is that both the supply and the demand sides are controllable despite the variability and intermittency of RE generation. The ESS of varying power capacities are indispensable in microgrids to maintain grid stability to avoid disturbances caused primarily by generation-demand mismatches. Higher penetration of variable RE into the distribution networks will urge the existing power utilities to evolve – in terms of intelligence and connectivity – in order to finally emerge as a system comprised of a fleet of power electronic components with distributed and embedded controllers for reliable operation.

Integration of RE sources into microgrids entails heterogeneous capabilities of its components and their controls like maximum power point tracking in solar and wind generators, controlled active-reactive power delivery, grid synchronization, grid formation, adherence to power quality as per the grid codes, etc. Yet, another challenge in microgrid management is the maintenance of continuity of power supply in cases of 100% RE generation. This is accomplished by smart selection of energy storage systems integrated at appropriate locations with appropriate size. And, the control capability of the storage has to targeted for bidirectional power flow, once for replenishing the charge under favorable grid conditions and otherwise for supporting the grid when needed. Further, control of ESS for transient power support will become a critical competency in microgrids to prevent the power variance of RE generators from affecting the system voltage and frequency profiles under short intermittencies.

Beyond the integration of RE sources and ESS to meet the local energy demand, the microgrid concept also includes conversion of *con*sumers to "*pro*sumers" and their involvement in energy management through various services and schemes like advanced metering infrastructure (AMI), demand response (DR), demand dispatch (DD), energy use scheduling, electric vehicle charging/discharging from/to the grid (V2G and G2V), etc. Realizing these in a microgrid requires distributed real time sensing, communication and control. Such capabilities, when added to microgrids through smart and connected embedded sensor nodes, convert the microgrids to smart microgrids (SMG).

The SMG is an ideal way to integrate the different RE sources at a community level and allows active consumer participation in the electricity business by enabling intelligence and communication capabilities. Furthermore, SMG is one of the most realistic paths to reach an 100% renewable energy grid with net zero emissions. Also, microgrids better manage peak loads – with local generation control, demand management, load shifting, virtual inertial support, etc., to improve the availability, affordability, quality, and security of supply.

1.2.1 MICROGRID DEVELOPMENT ACROSS THE GLOBE

The RE powered autonomous microgrids are a hot topic of research, development and demonstration with several working models currently in the international scenario. National governments of many developed countries profusely support such efforts financially.

The global market value of microgrids was recorded as close to 15 billion US dollars in 2017 and is envisaged to dominate the electric utility industry with a projected value of 30 billion US dollars by 2022. Such exponential growth is owing to the increased deployment in American and Asia-Pacific regions. North America held the largest market for microgrids till 2017 and is expected to lead the market in the future too, with the major share from the US. The Navigant Research's data on country-wise microgrid markets is presented in Figure 1.1.

However, it is expected that the worldwide market for microgrids will expand further due to improved grid resiliency and reliability, grid support during transient periods, climate change, and barriers on main grid expansion in remote locations, initiatives taken by various governments and optimization of grid assets and their management. Based on the region and the desired power sector development, one or more of these reasons will be the basis for initiation and installation of microgrids. The microgrid markets in the developed nations like the US and Japan often focus on providing supplementary support to their main grid in order to relieve its burden during peak demand and enhanced utilization of renewable sources. In contrast, microgrid markets in developing nations like India, Brazil, and China may focus on improving their electrification rates (in terms of reducing the size of their unelectrified population) and allowing for increased per capita energy consumption. For example, India has a total installed capacity of 290 MW, yet more than

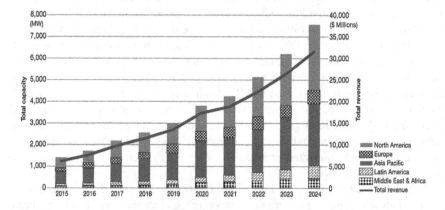

FIGURE 1.1 Microgrid Global Market Share.

250 million Indians have no access to electricity (even in the second decade of the twenty-first century) due to the prohibitive cost and technical as well as logistical challenges of grid extension to reach remote villages. The development of new grid infrastructure in under-developed nations like Africa would demand intensive capital and other resources which will challenge their economies.

State governments in developed countries like the US are beginning to liberally support microgrid efforts financially. Recently, the California Energy Commission offered huge grants for ten microgrid project proposals of about $51.9 million to accelerate the commercialization of microgrids in 2018. The winning projects propose demonstration of one or more of the major aspects of microgrids, such as resilience, clean power development, zero emissions, commercial scalability, energy security, and modularity.

1.2.2 MICROGRID INSTALLATIONS

There are several successful microgrid installations around the world with the largest currently located in the United States. A few of the worldwide microgrid installations are presented below broken out by country and highlighting their salient features.

- United States
  Fort Bragg Army Base, Fayetteville, NC, built the world's largest microgrid covering 160 square km, developed with the help of Honeywell. This microgrid showcases the reliability of power at reduced cost. It is established as a grid-connected microgrid working in synchronization with the military's electric utility. It has a central microgrid manager to monitor various generators which is fully integrated with the information technology and communications infrastructure. This microgrid has reduced the overall energy costs of Fort Bragg military base.

  California is one of the states in the US which has several microgrid initiatives that have resulted in many successful and visible deployments of microgrids. The microgrid R&D projects in the US include *University of California, San Diego's (UCSD)* campus microgrid, *Borrego Springs Microgrid* and the military microgrids of *Camp Pendleton* and *Fort Hunter Liggett*.

  The UCSD microgrid spreads across 450 hectares with 2×13.5 MW gas turbines, a 3 MW steam turbine and a 1.2 MW solar PV system meets almost 85% of the campus electricity demands. The UCSD uses a master controller called Paladin, which is responsible for the control of all the generators, storage and the loads with hourly computations and also optimizes the operating conditions of the microgrid. Paladin uses RE forecasting along with dynamic market pricing for the control of RE sources. Solar PV systems are provided with advanced controls like ramping up and down for obtaining optimum operation of the microgrid.

  The first commercial microgrid project implemented was the *Santa Rita Jail Microgrid* which is otherwise referred as the "Green Jail Project." This is developed with the objective of demonstrating reliability of an all

RE microgrid with large scale energy storage. It consists of 1.2 MW rooftop PV system, 1 MW molten carbonate fuel cell with CHP, 2 × 1.2 MW diesel generators for emergency, 2 kW wind turbines, and 2 MW Li-ion battery along with lighting and HVAC load. The system demonstrated dynamic islanding response with seamless disconnection and reconnections. It is jointly sponsored by Chevron Energy Solutions, Alameda County, Satcon Power Systems and Pacific Gas & Electric Company besides financial infusion from federal, state, and industrial funds.

- Japan

  Hachinohe Microgrid Project located in Aomori Prefecture, Japan was developed by the Regional Power Grid as an all RE microgrid funded by the New Energy and Industrial Technology Development Organization. It was in operation from October 2005 to March 2008, but became non-operational due to financial constraints. This project was a collaborative effort by the city of Hachinohe, Mitsubishi Research Institute and Mitsubishi Electric. This project had the objective to formulate and demonstrate optimum operation and control of a microgrid system in order to evaluate its power quality, operation cost and emission reduction. It consisted of 2 × 50 kW and 3 × 10 kW solar PV systems, small wind turbines, a 100 kW lead-acid battery bank and 3 × 170 kW gas engines driven by sewage and waste gas.

  The fully operational *Sendai Microgrid* achieved microgrid stardom because of its excellent performance during the 2011 earthquake and subsequent tsunami incidents which damaged the conventional electric utility facilities and washed away many local towns and villages. This microgrid comprising of natural gas fired turbo-generators, PV plant and modest battery storage is located within the campus of Tohuku Fukushi University.

- South America

  The microgrid in the remote Andes Mountains supplies the local *Huatacondo* community consisting of a diesel generator, solar plant, wind turbines and battery energy storage through a management system. This is a case of a remote microgrid where the local community takes ownership of operation and maintenance of the microgrid.

- Scotland

  The microgrid involving solar, wind and hydropower plants in kW capacities and installed on the *Isle of Eigg*, Scotland, claims a highly renewable power content aided by load management facilities with energy meters installed in all properties and battery storage employed for reliability of supply.

- Germany

  The microgrid project at *Mannheim-Wallstadt*, Germany, includes various DG units such as fuel cell, PV system, flywheel storage and combined heat and power (CHP) units addressing residential and commercial loads; it was developed to perform a swift and smooth switchover between grid-connected and islanded modes of operation.

- China
  Hangzhou Dianzi University's solar PV based microgrid coupled with capacitor and battery energy storage claims to be the world's first microgrid that employs 50% PV penetration. This microgrid is monitored and managed by a power control system with active power quality control.
- India
  In *Karnataka*, India, a solar PV with energy storage based remote microgrid has been deployed by the SELCO Foundation to provide electricity to Baikampady, a village near Mangalore. Also, remote DC microgrid projects are deployed in *Neelakantarayanagaddi Village, Mendare Village* and *Kalkeri Sangeet Vidyalaya*.

  A few other acclaimed microgrids in the world today are Illinois Institute of Technology Microgrid Project at Chicago, Maxwell Air Force Base in Alabama, the Kythnos Island Project near Athens, the Aperture Center in Mesa del Sol, Albuquerque, New Mexico, Ascension Island Microgrid project in Saint Helena, and the African microgrids at Annobon Island, Robben Island and Graciosa.

1.2.3 FINDINGS OF FIELD TRIALS

Investigations on microgrid operation generated elaborate reports and highlighted unresolved problems. Such experiences give clear directions for future microgrid research. Some of the issues reported are as follows:

- Maintenance of standard voltage and frequency at the prescribed levels of power quality is very hard when power balance is to be maintained with large penetration of RE sources in microgrids.
- The storage size needs increase with increase in the level of reliability demanded and thus storage systems occupy large spaces and require considerable maintenance in microgrids.
- Seamless transfer across grids when connected in autonomous modes of operation is difficult when continuity of supply is to be maintained at the highest quality level.
- Microgrid systems burden the main grid when they operate as a load.
- Implementing protection systems on microgrids is very challenging due to bidirectional power flow and related dynamics.
- Maintaining power quality and detecting harmonics are very tedious, as it is difficult to distinguish harmonics from the in rush currents of transformers and induction motors.
- Maintaining balanced three phase operation is very difficult owing to the presence of a large number of single phase loads of microgrid consumers and single phase generators such as solar PV.
- Energy storage management and comprehensive control with low latency communication are essential functionalities to be included in any microgrid.

1.2.4 ROLE OF POWER ELECTRONICS IN MICROGRID DEVELOPMENT

Power electronics (PE) is the important centerpiece in microgrid development and control. The PE converters are available in a wide range of specifications which are capable of carrying out various types of power conversions. Most of the RE sources and the energy storage systems are connected to grid/load through one or more power electronic interfaces. In solar PV and wind turbine systems, converters are controlled to extract maximum available power from the RE sources through appropriate tracking algorithms. These converters are either single stage or multistage depending on the nature of the generators' output. The DC-DC converters operated at high frequencies are commonly used for maximum power tracking. These DC-DC converters will serve as charge controllers in energy storage systems of microgrids, monitoring the state of charge of the battery besides performing the power dispatch control. In modern microgrids with AC and DC sub-networks, bidirectional AC-DC power converters are used to accomplish power exchange between the two sub-networks. The modern variable speed wind energy systems are connected to the grid through AC-DC-AC back-to-back converters which include two voltage source converters, one on the generator side and the other on the grid side working as rectifier and inverter respectively. Such switched converters on the generator side are controlled to achieve maximum power tracking, simultaneously mitigating the harmonics of the generator currents. At the same time there are PE converters provided on the grid side of RE generation systems which are controlled for their output voltage and frequency to establish and sustain synchronization with the grid. Generally, high power voltage source inverters (VSI) are common interfaces between the electric grid and RE sources. Controlled power delivery is accomplished through VSI by operating with closed loop controllers and pulse width modulation (PWM). Different PWM methods are adopted for converters ranging from the traditional sine PWM, space vector PWM, third harmonic injected PWM, etc. Advanced controls with implicit modulators are also expected to be developed for PE converters thanks to the availability of high speed digital controllers.

Miniaturization (imparting high power density) is a commendable feature that can lead to cost effective PE systems with smaller footprints. This feature can free up crucial panel space so that additional features and/or components can be realized without any further increase in the overall system size. High power density PE system design is still a challenge and it greatly depends on design trade-offs among the switching frequency, the size of filters and the thermal management system. Large switching frequencies of power converters will reduce the size of filters, but only at the cost of increased switching loss which in turn escalates the device temperature and creates a demand for larger heat sinks and cooling systems. On the other hand, when the converters are operated with small switching frequency, the filter size grows and cooling size shrinks. Thus, identifying an optimized PE system design is highly knotty, yet can contribute to high power density. The recent advances in power semiconductor technologies have substantially raised the maximum allowable junction temperature. The PE technology evolved over these years – in, current rating, switching speed and operating temperature – has improved the transient response too, significantly aiding the low inertia microgrids of the future.

The conversion efficiency of utility scale RE inverters is close to 96% but it drops considerably at high operating temperatures, say above 50°C. The new generation wide band gap (WBG) power electronic devices paved the way for achievement of high power density and low cost PE systems by operating at very high switching frequencies. Even at high operating frequencies robust high temperature performance of inverters is possible, thanks to higher allowable maximum junction temperatures of WBGs. This dual advantage will simultaneously decrease the size of both the passive components and the thermal management systems. Further, WBG devices are reported to have low on-state resistance that represents low conduction losses and higher efficiencies. So in the near future, PE converters with WBG devices will be more powerful, robust and energy efficient and can build more reliable and resilient microgrids which in turn can make electric utilities more robust.

Moreover, the modern high-speed digital signal processors and controllers complement the high-power high-frequency PE systems with their capability to process complex control algorithms within extremely short processing times. Thus the evolved cutting-edge PE system, that is extremely versatile and cost effective, can outstretch sophisticated control toward the highly integrated microgrid systems.

1.3   OPERATION AND MANAGEMENT

Integration of distributed RE-based generation, coordination of ESS and facilitation of demand side management in a large power grid need to happen at the microgrid level and in a distributed manner as the well-established, centralized larger grid should not be intercepted. Realization of energy management systems (EMS) which monitor and control energy production and consumption, in generation and consumer facilities, is a major development in power system research in the early years of twenty-first century. Decentralized energy management in microgrid systems is a multi-objective interdisciplinary problem addressing engineering, economic and environmental issues and offering enhanced flexibility aided with comprehensive analysis to ensure secure and reliable operation of the system. Performing energy management by introducing new tools on the DG is mandatory on the microgrid, as generation control in the conventional power plant is centralized and the former is distributed. This demands real time measured data communicated from the sensor network installed on the SMG. It can have automated operation through EMS by also interacting with the main grid.

Impact of microgrids on the stability of the main grid is evaluated through computation of frequency deviations, as this stability is a measure of power imbalance and must remain within operational limits. Once the main and microgrids are smarter, the deployment of EMS is effortless as the latter will monitor the status of the microgrid at regular intervals and take steps to maintain the real time grid frequency within the permissible limits. Failure of these efforts will cause serious damage to the appliances connected to the main grid and create a critical situation, potentially collapsing the entire system.

Several EMS for microgrids using different tools have been proposed and optimization models developed to implement the microgrid energy management by researchers worldwide. Most of these optimization models have used soft computing techniques based load forecasting. Various wireless communication methods – relevant for such applications, such as economic dispatch, optimized operation verified on laboratory scale, etc., are also available in the literature. Renewably energized base stations are able to trade shortage or surplus energy with the main grids to help grid-tied microgrid operations leverage their energy storage units using optimization techniques, minimizing the transaction cost and satisfying the worst case quality of service requirements.

Multi-agent modeling has also been used to realize operation of the EMS in microgrid. Software modular architectures were proposed and implemented for microgrid EMS, based on multi-agents that took care of additional services for advanced control. Multi-agent based control architectures were developed that can ensure robust, stable and optimal microgrid operation. Power trading among microgrids with DR and distributed storage was facilitated using agent-based EMS. Index-based incentive mechanisms were proposed to encourage customers to participate in DR. Performance indices based on electricity price, emission and service quality were developed and tested on grid-tied, islanded and multiple microgrids to test their performance.

Demand response is defined as a means of reducing the power consumption by end-users from their normal consumption pattern. These end-users respond to a price or reliability trigger from an independent system operator (ISO) thereby acting as *virtual power plants* and are encouraged to do so through either direct payment or incentives in their electricity bills. Introduction of DR programs among the end-users on the demand side is becoming an essential requirement to solve the global energy crisis. The major objective of DR is to reduce the power consumption during peak hours primarily performed by either load shedding or load shifting.

Two examples of microgrids employing DR are reported below:

- *Hartley Bay* in British Columbia, Canada, is a remote village housing a native community which relies on three diesel generators for electricity. The DR programs with appropriate load controllers are introduced in the commercial buildings to optimize the diesel dispatch. This is monitored using smart meters deployed to display energy use in real time while precision fuel flow sensors are installed on the generators to evaluate efficiency.
- San Diego Gas and Electric Company's utility microgrid in a residential community of *Borrego Springs*, California, explores the possibilities of price-driven DR via interaction with local consumer premises, electric vehicles, etc., using smart meters and home area network systems.

1.4   SMART DISTRIBUTION GRID

Advancements like integration of distributed RE generation and addition of ESS to the distribution network, distribution automation (DA), and modern demand side management (DSM) will completely revamp the style of operation and management of conventional distribution systems. Such renovated distribution systems will reflect the characteristics of a full scale power system at a micro level – each having its own localized generation, distribution and consumers – demonstrating the truest concepts of microgrids. Essentially, this process will demonstrate the evolution of the conventional distribution grid, to reflect the same principles as a microgrid or to itself be constructed from a collection of identical or diverse microgrids. The homogeneous features of a distribution grid formed by the identical microgrids or the heterogeneous features of one formed by diverse microgrids both demand the addition of smart, real time and networked devices to the distribution grid making it a smart distribution grid. The requirements, operational measures, targeted services, and capabilities of such a smart distribution grid will be completely different from those of a conventional grid.

A smart distribution grid offers a variety of services, applications and functions in realizing the vision of a smart grid. These include:

1. Islanding and networking of microgrids;
2. Smooth integration and control of RE sources, thereby realizing DG;
3. Wide area monitoring protection and control (WAMPAC) for real time monitoring and control of grid, realizing DA;
4. AMI, integrating consumers to the grid; and
5. DR, DD, Net Metering, G2V, V2G, etc., together realizing modern DSM.

Any smart distribution grid will have two phases of operation: (i) Synchronized operation with the main grid in the grid-connected mode of microgrids, and (ii) stable and balanced operation of microgrids in the islanded mode. The requirement in the grid-connected mode of operation of all associated microgrids is to do real time monitoring of grid parameters and follow instructions from the control centers of the main grid for stable and balanced operation. In the islanded mode of operation of the microgrids, the real time monitoring and control have to happen as per instructions from the respective microgrid control center.

Whether islanded or grid-connected, the various features or applications of smart operation have to be carried out in real time. This decision making is partly centralized and partly distributed. All other actions in grid-connected mode, except protection which is done locally, are executed in the microgrids as per directions from the appropriate control center. Decision making happens in islanded mode, too, in the same way, but all these are local to the microgrid. For these decision making processes, distributed and real time data are to be communicated from various field devices to the appropriate higher level devices and control centers and the control commands must be communicated back. Therefore, there is a need to enable synchronous operation of a large number of geographically distributed individual entities and it demands the distribution grid to be smart. This is achieved by integration of various intelligent and connected embedded sensor nodes throughout the distribution grid. This adds a communication layer over the power system network bringing in the capability of bidirectional communication enabling automation at various levels of operation of the smart distribution grid.

BIBLIOGRAPHY

1. www.galvinpower.org/resources/microgrid-hub/smart-microgrids-faq/examples
2. www.power-technology.com/comment/global-microgrids-market-to-reach-30bn-in-2022/
3. Alireza Aram, Global Innovation Report-Microgrid Market in the USA, www.hitachi.com/rev/archive/2017/r2017_05/Global/index.html
4. http://microgridprojects.com/india-microgrids/
5. Wood, E. "California Names 10 Winners for $51.9 Million in Microgrid Grants," Available online: https://microgridknowledge.com/microgrid-grants/, accessed February 21, 2018.
6. "Energy access database," International Energy Agency, Available online: www.iea.org/energyaccess/database/, accessed on February 3, 2020.
7. "Access to electricity—% of population," World Bank, Sustainable Energy for All (SE4ALL) database. Available online: https://data.worldbank.org/indicator/eg.elc.accs.zs, accessed on May 30, 2019.
8. https://china.lbl.gov/sites/all/files/lbnl-5825e.pdf
9. Saeed, S. "Role of Power Electronics in Grid Modernization, Power Technology Research," Available online: https://powertechresearch.com/role-of-power-electronics-in-grid-modernization/
10. Kizilyalli, I. C., Carlson, E. P., Cunningham, D. W., Manser, J. S., Xu, Y. A., and Liu, A. Y. "Wide Band-Gap Semiconductor Based Power Electronics for Energy Efficiency," Available online: https://arpa-e.energy.gov/sites/default/files/documents/files/ARPA-E_Power_Electronics_Paper-April2018.pdf, accessed on March 13, 2018.

2 Design of Microgrids

A. Vijayakumari

CONTENTS

2.1 DEFINITION, STRUCTURE AND COMPONENTS

2.1.1 EVOLUTION OF MICROGRIDS

The maiden power system developed by Edison with Pearl Street Power Station, Manhattan utilized DC power. Constrained by short transmission distances and large conductor size, DC power gave way to Tesla's AC technology that prevailed. Yet, it was imperative for Edison to maintain his small power plants to serve local customers in city centers within a span of about 1.5 km. In fact, this operational need is the outset of the present-day concept of the microgrid. However, George Westinghouse's Niagara Falls Project in 1895 realized long distance transmission of high voltage AC power through about 32 km (Figures 2.1 and 2.2). The early electricity consumers were predominantly with Edison's power company and were all using only DC loads. There were research efforts to obtain high voltage DC through series of motor-generator sets, but the prohibitive cost involved in the rotating machines and their intensive maintenance made high voltage DC an economically unattractive choice. Gradual acceptance of AC power by industrial consumers unleashed apprehensions about the DC technology. This duality of available options led to both DC and AC coexisting in many consumer premises with changeover switches and accessories. Thus, the hybrid electric utility – foreseen today to be the next generation hybrid microgrids – can be traced back in history as the coexisted DC-AC systems of Edison's and Tesla's era. The decline of DC happened gradually, and the electric utility emerged as today's synchronous AC networks with centralized power generation and delivery. It has been proven that AC and DC can coexist while each serving its purpose according to the characteristics and demand of the load. Of late, this synchronous electric network is surprisingly metamorphosing toward the *War of*

FIGURE 2.1   750 kW GE make alternator Folsom hydroelectric power plant in California, installed in 1895. (Photographed on November 2, 2019.)

FIGURE 2.2   Synchronizing panel in Folsom power house. (Photographed on November 2, 2019.)

Current period complexities like decentralized generation, coexistence of AC-DC networks, coexistence of low- and high-voltage levels in a common network, coexistence of AC and DC loads, decentralized control, integration of diverse generation mix, etc. Yet, the AC synchronous network supports as a backbone to all these reformations in generation dispatch and demand management.

Utilities across the globe are yearning for the day when there will be 100% renewable energy (RE) penetration into the electric utility system. But, large scale integration of variable RE may raise the complexity in grid operation and management several fold and may even lead to disruption of the synchronous network. However, this can be addressed through addition of large energy storage systems onto the electric grid in order to buffer the variability of RE generation. If RE is utilized in microgrid, a small island of electric network located downstream on the distribution grid, it can be a prospective way to gain high RE penetration supported by addition of reserves in the form of energy storage. Such autonomous energy systems with high penetration of RE require affordable and reliable power management strategies – including features such as, definitive storage systems, highly interconnected control strategies, flexible generation dispatch and demand side management – to balance the system dynamics.

In short, a redesign of the power system topology is necessary to accomplish the 100% RE vision. With the evolving fleet of new generation loads and with the legacy industry load (which uses DC at intermediate levels in their power delivery chain), the share of DC power demand is increasing day by day. Therefore, the availability of DC power can eliminate the AC-DC conversion losses in the majority of the electronic loads of today starting from computers, medical equipment, smartphones, infotainment electronics and lighting. Even for the variable speed motor drives which include air conditioners, industrial drives and electric vehicles, DC is the right source of power. Moreover, the power quality concerns in conventional AC systems, owing in large part to the increased deployment of AC to DC conversion units, can be abated by avoiding all such conversion on AC grid and shifting the respective appliances to the DC grid. Hence an innovative microgrid topology should tend to eliminate most of the intermediate power conversions which are indispensable with the legacy AC grid. The imminent microgrid is expected to satisfy the following salient capabilities:

- support the existing power sources for the traditional AC loads,
- be capable of simultaneously supplying DC power,
- be able to utilize a diverse mix of local RE sources,
- include energy storage as an obligatory component to buffer the variability of RE,
- optimally locate RE and energy storage in order to reduce the system losses, and
- eliminate the need for long distance high voltage transmission lines.

Further, there is a need to standardize the topology of microgrids, identify the essential power converter interfaces and develop the right control strategies for reliable operation of the whole system in tune with the AC grid whenever needed.

2.1.2 CONCEPT OF MICROGRID

Reliability and quality of power are major concerns for today's electric utilities owing to the ever-increasing supply-demand gap and increased penetration of non-linear loads. Increased electrical system efficiency, reduced transmission congestion, competitive electricity pricing and grid independence for individual consumers are some important features envisaged in the near future for electric utilities. A single, yet comprehensive, solution for such multidimensional demands on these utilities is the microgrid.

A microgrid is an interconnected electric energy system that has its own generation and loads and can operate while connected parallel to a major electric grid or in an islanded condition. So, microgrids are treated as dispatchable loads, as local loads – fed by local generation through local controls – microgrids can relieve the main grid of considerable burden. Microgrids are envisaged as the enabling technology to make the present-day grid *future ready*. It allows large consumers to utilize their electricity supply efficiently and economically with the help of local generating assets like micro-generators, renewable energy generation, etc. Consumer wait time for their utility to extend infrastructure and augment generating capacity can be substantially reduced, and that too at a lesser cost. Smart microgrids set the scene for innovative future electricity utilization and corresponding policies, so that future consumers will not be just electricity users, rather they will be smart prosumers and smart energy resources.

2.1.3 Definitions

Microgrids are a group of interconnected generators, loads and storage with an energy management system for local energy delivery that exactly meets the possible local demand at any given point of time. Such accurate energy management is possible because of the design of its entities for the constituents being served, at the planning stage itself. The varieties of constituents of microgrids range from a village, a city, a commercial complex, hospitals, educational institutions, etc.

Various international agencies have come up with definitions describing the entities and features of microgrids at their best. Some of the definitions are as follows:

IEEE Standards Association: Microgrids are localized grids that can disconnect from the traditional grid to operate autonomously. Because they are able to operate while the main grid is down, microgrids can strengthen grid resilience and help mitigate grid disturbances as well as function as a grid resource for faster system response and recovery.

US Department of Energy: A microgrid is a group of interconnected loads and distributed energy resources within clearly defined electrical boundaries that acts as a single controllable entity with respect to the grid. A microgrid can connect and disconnect from the grid to enable it to operate in both grid-connected and island mode.

Microgrid Institute: A microgrid is a small energy system capable of balancing captive supply and demand resources to maintain stable service within a defined boundary. There's no universally accepted minimum or maximum size for a microgrid.

CIGRÉ: A microgrid is an electricity distribution system containing loads and distributed energy resources (such as generators, storage devices or loads) that can be operated in a controlled, coordinated way while islanded from any utility grid, and is under the control of a single management entity.

ASEAN Center for Energy: The microgrid system is a small power supply system that consists of loads and distributed energy resources, such as renewable energy sources, cogeneration, combined heat and power (CHP) generation, fuel cell and energy storage systems.

2.1.4 Characteristics of Microgrids

There are distinctive characteristics which differentiate microgrids from that of distributed generators – these two are often confused. The basic block diagram of a typical microgrid network is presented in Figure 2.3. It has various generators like RE (wind and solar) and fossil fuels (gas and diesel), energy storage devices and a collection of loads as shown. The microgrid system is connected to the distribution feeder

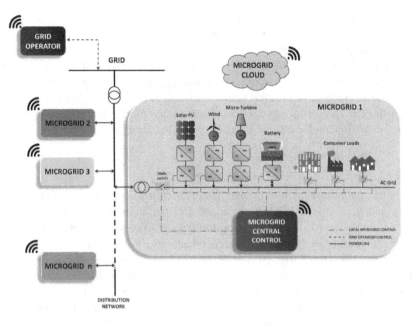

FIGURE 2.3 A typical microgrid-connected to legacy grid.

through a static switch at the point of common coupling. Most of the generators/ energy storage system are connected to the microgrid through power electronic converters and their respective source controllers. There is a master controller which usually exchanges the current operational data with the source controllers and takes care of the overall network control, decision making and sometimes the generation forecasting, too. The microgrid has two operating modes, viz., the grid-connected mode and the islanded or autonomous mode. In the grid-connected mode, the generators are controlled to deliver their respective maximum powers, and if it exceeds the total microgrid demand at that moment, then the excess power is exported to the utility. So the generation controllers are assigned power references as dictated by the respective maximum power point tracking (MPPT) algorithms of RE sources. However in autonomous mode, the central controller assigns the required reference voltage and frequency for the source controller to establish the grid. Power references are assigned to the generation controllers depending on the local demand information. The control enables *plug and play* of any source at any given instant of time.

2.1.5 BENEFITS AND CHALLENGES OF MICROGRIDS

The electric utilities are expected to undergo rapid progress due to fast integration of large shares of various RE sources. The optimal way to achieve this is by developing microgrids, as the microgrids can manage the power variability of RE locally, leaving the operation and stability of the regional grids unchallenged. Such utility transitions certainly bring encouraging benefits to both the utility company as well as the consumer. For the consumers, it can provide utility services of high standard, reliable electricity supply, attractive tariff rates, etc. The future utilities give way to consumer participation in grid management through concepts like demand response, demand dispatch and power purchase from prosumers, etc. The utility benefits are far more than those felt by the consumers, viz., reduced network congestion, effective peak load management, reduced line losses, optimum generation mix with energy storage, less generation capacity addition than the increased demand, higher plant load factor on all generators, possible deferral of infrastructure investment, ease in ancillary service provision, and above all, clean energy integration.

2.1.5.1   Benefits

2.1.5.1.1   Reliable and Resilient Electricity Supply

The most important benefit of microgrids is the reliability it provides to the electric utility and its customers. When the main grid begins to fail, the microgrids can still remain active to support their customers by disconnecting or islanding from the central grid. Microgrids being independent entities in terms of local generation and storage, they can serve the customers until the main grid supply is restored. Energy resiliency is a term closely related to reliability, and often confused with it too. *Reliability* is the ability of the utility to keep the power on for its customers, and *resilience* is the ability to circumvent power outages and/or to revive quickly on loss of generation under any such unforeseen circumstances like accidents or natural calamities. The microgrids are designed and automated to restore essential services rapidly, even under sudden disruptive circumstances.

2.1.5.1.2   Reduced Transmission Losses

The benchmark of average utility line loss is in the range of 6%–10%, but in several countries these losses are at alarmingly high levels due to the geographic and economic constraints. Heavy line loading being the prime reason for high transmission and distribution (T&D) losses, the line losses can be reduced to a great extent by restricting peak hour demand. As microgrids have local generation to cater to the local loads, the demand on the main grid and the related T&D losses can be substantially reduced. Local compensation of reactive power on the microgrid can reduce the line loading further. Additionally, the distributed generation reduces the need for long transmission lines to some extent and thus helps in the reduction of T&D losses too.

There are various non-technical losses in the electric power supply system like theft, pilferage and losses due to poor metering, ineffective billing, absence of energy auditing procedures, and sporadic maintenance of equipment. Nearly all of these non-technical losses can be eliminated through the determinative solution of a smart microgrid as it includes distribution automation with smart meters, dynamic electricity pricing, demand response programs and ICT enabled billing, collection and accounting.

2.1.5.1.3   Reduction in System Capacity

Rapid growth in demand forces several pieces of equipment and feeders in the utility system to operate at rated capacity. This calls for upgrade of transmission lines, transformers and other equipment which involve high investment cost besides right of way and other legal concerns. However, if the utility is to meet the peak load, then large capacity has to be built yet it must be maintained at lower plant load factor which is economically not attractive. Often, such feeder extensions are through remote locations and are time consuming and far more difficult to maintain. Allowing microgrids penetration in the distribution system, any peak demand can be locally catered to. This relieves the stress on both the transmission and distribution lines. Thus, neither the central generation capacity, nor the distribution feeder capacities need to be supplemented, as the microgrid solution is part of the load centers. Microgrid as an alternative is more economical than the traditional solutions because it defers the cost of capacity addition with attractive complimentary benefits like improved reliability, peak demand reduction and easy and low cost maintenance.

2.1.5.1.4   Integration of RE
Increased awareness of clean energy and its economic benefits promoted large investments in RE development in the recent past. But, large scale integration of RE may disturb the normal operation and management of the synchronous grid owing to the intermittency and variance in such power generation. Frequency and voltage fluctuations are some of the serious concerns on the part of the utilities regarding large scale RE penetration. On the contrary, if large numbers of small scale RE generators are allowed to penetrate at the distribution level, then the dual advantage of high RE share and less risk in energy management can be achieved. One or more small RE generators with reliable power management strategy and energy storage forms a microgrid. Microgrids can thus be instrumental in achieving the goal of 100% RE penetration into the electric utility.

2.1.5.1.5   Cost of Reliability
The *cost of reliability* is the cost that would be involved in transforming the present-day utility electric networks to accomplish reliability levels on par with these future smart microgrids. It refers to an attribute workable only if the legacy grid is appended with fast communication infrastructure, sophisticated controls, sensors, automatic black start capability after a fault and system health monitoring which are not part of it today. Similarly, provision for receiving weather forecast data and utilizing it in planning and scheduling the ramp rates of other generators are inherent features in microgrids that are not available in legacy grids. Infrastructure required to achieve these features in legacy grids is massive, and cost is formidable, due to the geographic spread of conventional systems, whereas the cost of reliability in microgrids is much lower as the generator controls and the load management systems are present locally.

2.1.5.2   Challenges
Balancing out the widespread benefits of microgrids, there are multifaceted challenges – ranging from technical to economic, from policy matters to social concerns. The non-technical challenges include issues related to policy and ownership and lack of regulatory suggestions and business models. However, the technical issues are of primary consideration, as their resolution is key to accelerated diffusion of microgrids into current electric networks.

The technical challenges are associated mainly with the planning and modeling, control characteristics and response of protection units. These challenges are imposed on the same microgrid components that operate in the grid-connected as well as in the islanded mode. But the system behavior and the control demands are not the same under these two modes of operation. For example, islanding detection and resynchronization are the main capabilities when working in the grid-tied mode, whereas frequency regulation and power management are vital in the autonomous operation. Another major challenge is the microgrid protection system that has to respond to the main grid as well as the microgrid faults with two different levels of fault currents. A few technical challenges are discussed in detail in the following sections:

2.1.5.2.1 Stability of Microgrids

The microgrid stability is to be dealt with differently for grid-connected and islanded conditions. The grid-connected condition poses very little challenge to microgrid stability as the grid extends the power support for instantaneous balance. But, the autonomous mode with variable generation by wind and solar combined with varying load can create instability. Microgrids in autonomous mode with high RE penetration require reliable power management strategy including definitive storage system, highly interconnected controls, precise load sharing, flexible generation dispatch and demand side management to sustain stable operation. Reactive power compensation and fault voltage ride through capability are essential in autonomous mode in order to ensure voltage stability. Further, microgrids should be designed for optimum generation mix, which can give a higher degree of flexibility required for enhanced stability management.

2.1.5.2.2 Microgrid Control

Every microgrid requires a master control system with multi-layered software compartments to embed the entire control, monitoring and data acquisition tasks. There are various tasks including generation and demand control, resource and load forecasting, sensor data acquisition and decision-making algorithms, grid monitoring and market tariffs, power balance and frequency regulation, effective utilization of RE resources, battery management systems, GUIs and many more. Identifying such a system remains a challenge as it has to have massive storage and computing capabilities with different response times for different tasks and modes of operation. Another major concern is selection of the type of microgrid control – centralized control, distributed control, hierarchical control or coordinated control. There is no consensus at this time which control system will optimize system behavior.

2.1.5.2.3 Protection in Microgrid

An appropriate protection system for microgrid should be provided such that it responds for the faults within the microgrid and faults on the feeder where it is connected. When the faults are within, then the respective fault has to be isolated which may be result in several sub-micro islands. Then it will be necessary to take care of the reliability of supply within these sub-micro islands. When the fault is on the feeder, then the protection system has to isolate the microgrid from the utility within a short response time to protect the components of the microgrid. The sensitivity of the protection system has to ensure that there are no undetected faults or delayed trips. The selectivity of the protection system should be designed in such a way that no false tripping happens under any operating condition. Even under the circumstances where the fault current magnitudes are at par with the load currents of the microgrids, the protection system should isolate it from the feeder. Also, with low magnitudes of internal fault currents, the protection system should respond and isolate that small portion of the microgrid. An adaptive microgrid protection system with proper communication to dynamically vary relay settings is still a challenge.

2.1.5.2.4 Island Detection in Microgrids

Islanding is a condition during which the electric utility to which the microgrid is connected is absent for whatever reason. Islanding can be of two types – unintentional islanding and intentional islanding. The first is primarily due to the electric utility shutting down for maintenance, load shedding or fault clearing. Intentional islanding is executed by the microgrid management in order to provide higher power quality and efficiency for local customers/loads.

The earlier grid protocols (IEEE 929-2000, IEC 62116, and IEEE 1547) recommend that any type of distributed generation must shut down during times of absence of mains. But the protocols have now changed as microgrids can still operate in the autonomous mode even without the utility. Islanding detection is one of the challenges in microgrid implementation, as faster and accurate detection methods are necessary to avoid power outage within the microgrids and to ensure continuity and stability of service. Islanding detection methods are broadly categorized as passive and active methods. The former uses transient behaviour of voltage, current and frequency to detect mains failure, while the latter periodically injects a non-characteristic signal and detects the absence of mains by analyzing its response using various signal processing techniques. Selecting a suitable islanding algorithm is based on the response time, impact on power quality and the presence of non-detection zones (NDZ). The active methods have proven to have small NDZ, but they suffer from large response time due to laborious computations and power quality issues resulting from signal injections. In spite of quick response and no harmonic footprints, passive methods suffer from large NDZ which is far more hazardous for the microgrid equipment than feeder faults. Islanding detection is still an open-ended challenge for microgrid implementation.

Besides these, there are several additional challenges such as control for seamless mode transfer, location, size and type of generation and storage equipment in microgrids, logic for power sharing, microgrid planning, energy management under different modes, metering and tariff, load management and demand response, coordination of interconnected microgrids, cyber security, regulatory policy framework, etc. Rigorous research efforts are ongoing in terms of demonstration units and scalable prototypes in all these domains envisaging enhanced and optimized microgrid performance and evolving the smart microgrid.

2.2 STRUCTURE, COMPONENTS, AND OPERATION OF MICROGRIDS

2.2.1 CLASSIFICATION OF MICROGRIDS

Though *microgrid* is a universal term representing a localized group consisting of energy sources and interconnected loads, they can be distinguished from one another based on the power supply, location and structure. Microgrids often have

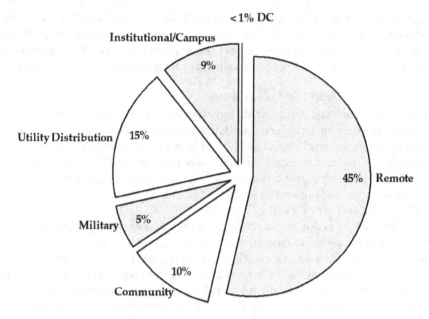

FIGURE 2.4 Total microgrid power capacity market share by segment, world markets: 2Q 2019. (From Navigant Research.)

cross correlation across these types such that individual units can be identified under more than one classification. As an example, the percentage of shares of different segments in the power capacity market is presented in Figure 2.4.

2.2.1.1 Classification Based on Power Supply

According to the power supply available for the loads within the microgrid, it can be classified as either an AC microgrid, DC microgrid or hybrid microgrid.

2.2.1.1.1 AC Microgrids

AC microgrids are very *similar to* the legacy grid except that these will be powered by local generation, controlled through local controllers and cater to local loads. Almost all the power sources are connected to the AC network through power electronic converters. The voltage and frequency regulation will be entrusted to one of the converters in the local network. A typical AC microgrid is shown in Figure 2.3. One of the major limitations of this type of microgrid is the additional conversion required for DC power. The outputs of major micro sources being DC, these need multistage converters with synchronization capability for sustained operation on AC network. The fast-emerging DC loads like LED lights, infotainment systems and battery chargers of electronic gadgets obtain power through multiple AC-DC conversions, thus reducing the efficiency of the system. However, it is possible to directly integrate all the micro sources with minimum network modifications and using conventional grid following inverters. Transformers can be used to achieve different voltage levels required for the loads. The AC microgrids when operated in islanded condition need reactive power support, making it necessary to have reactive power sources within the local network.

2.2.1.1.2 DC Microgrids

Solar PV generators integrated with BES offer a great opportunity for DC grids of any scale. DC microgrids can potentially be deployed anywhere, as solar radiation is omnipresent on the globe. Educational institutions, industrial facilities and residential apartments can have their lights (LED) and fans (Brushless DC Motor) as DC loads, besides mobile phone charger interfaces. DC loads along with Solar PV and BES are likely to be accommodated together to form dedicated DC grids. Unlike in AC grids, no reactive power support is required in DC network. The traditional AC loads need to be operated through inverters. However, this arrangement will be economically prohibitive with high power AC loads. A total grid restructuring would be required on the existing distribution network to accommodate such modification in power supply. A typical DC microgrid in depicted in Figure 2.5.

2.2.1.1.3 Hybrid AC-DC Microgrids

This configuration combines the advantages of both AC and DC network architectures. It has AC and DC sub-grids with generators and loads on either side. Direct integration of both AC and DC based micro sources, BES and heterogeneous loads is thus made possible. A typical hybrid AC-DC microgrid is depicted in Figure 2.6. Minimized conversion loss between AC and DC networks with reduced interfacing components also ensures increased reliability, improved system efficiency and reduced costs. Power exchange between the AC and DC networks will be facilitated as needed through bidirectional AC-DC converters. Such exchanges ensure continuity of supply on both networks by sharing

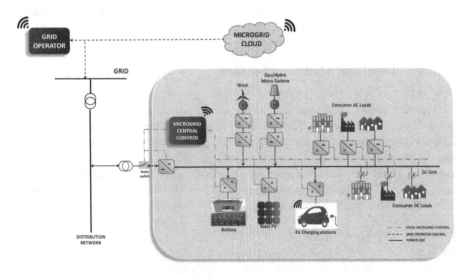

FIGURE 2.5 Structure of a typical DC microgrid.

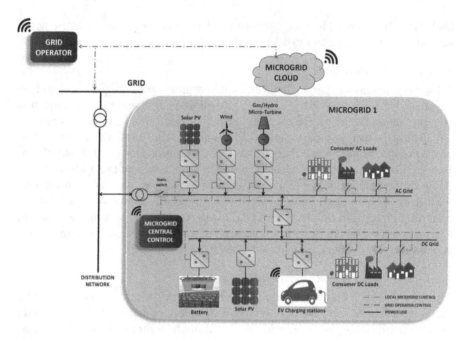

FIGURE 2.6 Structure of a typical AC-DC hybrid microgrid.

the excess generation on either side. It also helps to store excess AC generation in BES connected on the DC network as well as to extract BES stored energy to meet any deficit on the AC network. Further, the power exchange can ensure maximum utilization of RE sources by diverting the available generation to the needy locations or to the storage. On the consumer end, hybrid microgrids offer the flexibility to choose between AC and DC as required by their equipment. Minimum modification on the present AC distribution network will be sufficient to establish such hybrid systems.

2.2.1.2 Classification Based on Location

2.2.1.2.1 Community or Campus Microgrid

Universities, schools, commercial complexes, military camps, corporate offices, etc., are well suited to be served by community microgrids. This type of microgrid can have a grid connection which serves as a backup power option. Beyond which fossil fuel–based generation can be kept as an emergency backup. The local community can take care of the maintenance of the microgrid assets and smart communication features can be utilized for its monitoring and control. Such information and communication infrastructure and their capabilities are presented in Chapter 3.

2.2.1.2.2 Industrial Microgrid

An industrial microgrid offers power supply security and reliability to industrial consumers like automated manufacturing and processing plants. In such premises, loss of power supply may result in severe revenue loss and further require long resumption time of the processes when power is restored. An appropriate generation mix including fossil fuel with adequate storage and fast acting controllers can bring additional advantages like secure power supply, reduced running cost, obligatory RE utilization and energy efficiency benchmark attainment. These microgrids are mainly intended as captive generation systems.

2.2.1.2.3 Stand-Alone or Off-Grid Microgrid

Sites such as islands, grid-deprived remote locations like tribal hamlets and isolated colonies can opt for stand-alone microgrids which are not connected to any local electric utility. Such microgrids are to be provided with black start capabilities to allow them to start generation at any required point.

2.2.1.2.4 Utility Microgrid

Utility microgrids are connected to the main grid and can exchange energy with the main grid at the point of common coupling. Such microgrids include a distribution feeder and one or more distribution substations within its spread. They may have several generators of different types and connected loads of various capacities. A distributed grid management system has to be in place to monitor and control the energy exchange within the grid at various segments. An islanding detection unit is mandatory as this can work both in the autonomous mode as well as in the grid-connected mode. Seamless mode change is targeted upon an outage. Additional operational features such as grid-forming operation, distribution state estimation and load flow calculation are often required for these microgrids.

2.2.1.3 Classification Based on Geographical Spread and Capacity

Based on their geographical spread and capacity, microgrids are classified as mini-grid, microgrid and nanogrid. World Bank defines minigrid as an "Isolated, small-scale distribution network typically operating below 11 kV that provides power to a localized group of customers and produce electricity from small generators, potentially coupled with energy storage," while Lawrence Berkeley National Laboratory defines nanogrid as, "A small electrical domain connected to the grid of no greater than 100 kW and limited to a single building structure or primary load or a network of off-grid loads not exceeding 5 kW, both categories representing devices capable of islanding and/or energy self-sufficiency through some level of intelligent distributed energy resources management or controls."

The terms mini-, micro- and nanogrids represent the size, capacity and configuration of both the off-grid or grid-connected systems. Nanogrid is the smallest discrete network unit with the capability to operate independently like a building-level circuit with building-integrated power generation source(s).

2.2.2 Components of 100% RE Microgrids

Diverse ranges of distributed energy resource exist for microgrid deployment which includes diesel generator, micro gas turbine generator, fuel cell, solar photovoltaic (PV) array, wind turbine generator (WTG) and micro hydroelectric generator. Similarly, a wide range of energy storage options is also available like flywheel, varieties of battery, super capacitor, etc. The specific mix of generators and ESS has to be judicially selected to match project specific needs like the load profile, microgrid location and the economic constraints. The energy storage capacity will be decided based on the variability of RE generation in a given location and the local demand profile. Energy storage system design will be targeted to make it act as a buffer in balancing the generation with the demand on the microgrid. Appropriately chosen and sized storage can make the RE generation dispatchable and controllable. Additionally, the storage capacity can be marginally augmented in order to extend ancillary grid services like virtual inertia support, voltage control support, peak shaving and valley filling. Thermal and mechanical energy storage can also be added wherever appropriate.

When 100% RE is targeted, then the generation options are limited to PV, small WTG and micro hydro generators (including pumped storage options). The general architecture of RE microgrids can be classified broadly with respect to control functionality as autonomous or grid-integrated systems.

2.2.2.1 Components of Grid-Integrated Microgrids

The terminal voltage-current characteristic of most RE generators are distinctive and do not match with the characteristics desired for grid-feeding. Power electronics has therefore become an integral part of microgrids with technologies like PV, variable speed WTG and battery which need power electronic interfaces with heterogeneous controls for effective transfer of power to the network. Depending on the type, whether it is DC or AC, the converters are selected from among DC-DC, DC-AC, AC-DC, bidirectional DC-DC, etc. In grid-integrated DC microgrids, either single stage or multistage DC-DC converters are interfaced between the PV array and the electrical network. When the microgrid operates in the grid-integrated mode, maximization of power extraction from each source is often coordinated through maximum power point tracking (MPPT) as shown in Figures 2.7 and 2.8. In the case of

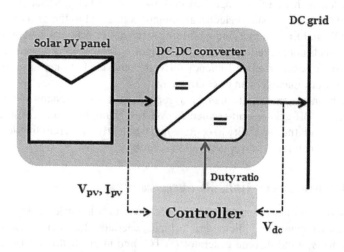

FIGURE 2.7   PV connected to DC grid.

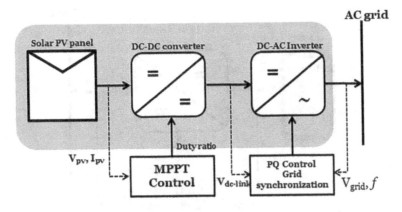

FIGURE 2.8 PV connected to AC grid.

the DC system, the converter is controlled to operate in parallel to the DC grid, simultaneously delivering maximum power extracted from the PV array. An inverter is interfaced between the DC-DC converter and the grid in the case of the AC system. Grid synchronization and power control are the responsibilities entrusted with the grid-side inverter control. The grid information like DC voltage, PV voltage and PV current are utilized for converter control in DC grids. On the other hand, the DC link voltage, grid voltage and frequency (besides the PV metrics) are required for synchronized power control in AC grids. Smooth islanding, soft start/reconnect and resynchronization with the main grid are the additional control features required for converter control.

The energy storage offsets the disturbances caused by fluctuating RE generation on one side and fast load changes on the other and maintains the stability of the microgrid. Battery is connected to DC grids through a bidirectional DC-DC converter enabled with a charge-discharge control feature as shown in Figures 2.9 and 2.10. The charge-discharge control has to protect the battery from over charge as well as deep discharge, too. The battery can provide one or more of the aforesaid ancillary services to the grid when needed, so the controller should be capable of receiving such command signals from the microgrid controllers. The battery controller in a DC grid will target to provide transient power support to maintain the power output constant irrespective of the RE variability and operate in parallel with other converters as shown in Figure 2.9. In an AC microgrid like the one in Figure 2.10, the synchronized grid-side converter will be controlled to operate in either inverter or rectifier mode as demanded by the operating conditions.

Among the several WTG options available, the variable speed ones like doubly fed induction generator (DFIG) and permanent magnet synchronous generator (PMSG) are suitable for microgrids, as these have higher energy conversion efficiency and can be operated in either islanded or autonomous mode as desired in microgrids. The DFIG can be operated with slip power rated AC-DC-AC converters on the rotor circuit with the stator directly feeding power to the AC grid as seen in Figure 2.11. Rotor power at slip frequency is first converted to DC and then

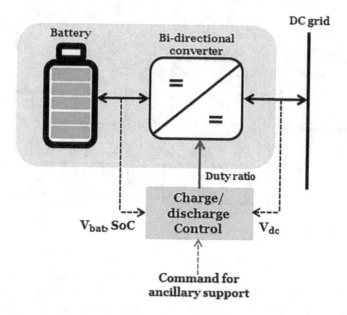

FIGURE 2.9   BES connected to DC grid.

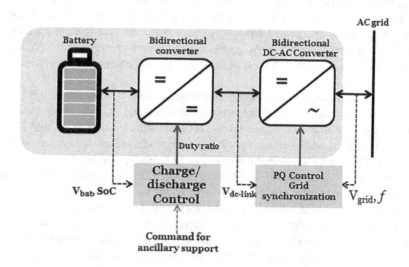

FIGURE 2.10   BES connected to AC grid.

back to AC at grid frequency. The rotor power can be in either direction – outward or inward, depending on whether the rotor speed is sub-synchronous or super-synchronous. The rotor power can be either sent/received to/from the grid through the grid-side converter as seen in Figure 2.11. Alternatively, the rotor power can be directed/retrieved to/from a battery as in the scheme of Figure 2.12. A battery being a mandatory component in microgrids the second scheme would be a smart choice

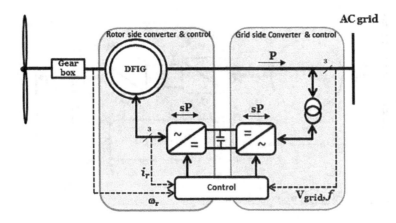

FIGURE 2.11   Grid tied DFIG based WTG.

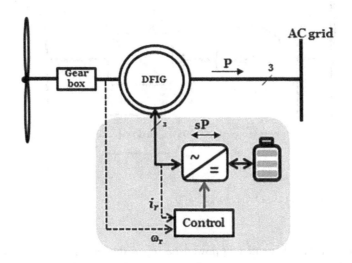

FIGURE 2.12   Grid tied DFIG with battery support.

which eliminates the grid-side converter and its synchronized control thus simplifying the scheme. Operating DFIG in DC microgrid is practically complex because it needs, in addition to the rotor side converters, a full-rated AC-DC converter on the stator along with reactive power support.

The PMSG is another candidate of variable speed WTG with an inherent feature to operate on DC grids through a single stage AC-DC converter. The PMSG of Figure 2.13 has two full-rated back-to-back AC-DC-AC converters to connect to the AC microgrid. The machine side converter can even be an uncontrolled rectifier to reduce the control complexity. The grid-side converter control features are synchronization, power control and sometimes MPPT. For PMSG in DC microgrids, the controlled AC-DC converter should be operated in parallel with the grid and it often

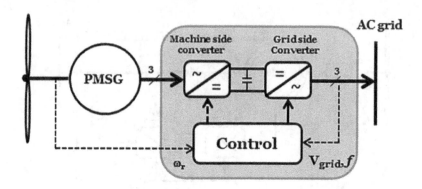

FIGURE 2.13   Grid tied PMSG based WTG.

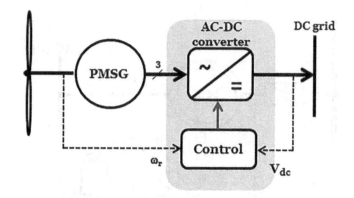

FIGURE 2.14   PMSG based WTG on DC grid.

has droop control implemented for power sharing as shown in Figure 2.14. When the machine side converter is a diode rectifier, then another DC-DC converter is used in series with it (Figure 2.14).

2.2.2.2   Components of Autonomous Microgrids

The components of both DC and AC autonomous microgrids are essentially similar to those of the grid-connected system, except that the control functionalities differ as there is no synchronization; instead, autonomous microgrids need black start capability. The converters in the autonomous systems should have a control feature to establish the respective system with the rated voltage, either DC or AC, and should regulate the voltage. Frequency regulation also will be entrusted on converter controllers in AC systems to establish the rated frequency of the system. Solar PV fed DC and AC autonomous microgrids are shown in Figure 2.15a and b.

Similarly, the variable speed DFIG can be operated in stand-alone mode through rotor side control. The DFIG scheme with battery at the rotor circuit can provide the

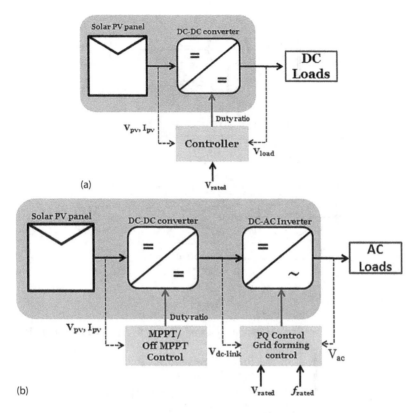

FIGURE 2.15   Solar PV fed (a) DC and (b) AC autonomous microgrids.

black start capability to the microgrid. That means, this microgrid can start operation with the battery power injected to the rotor and in turn can make the generator establish and sustain an AC grid independently. A common feature across all the autonomous systems is that a reference voltage and frequency is predefined in their control as seen in Figures 2.15b and 2.16a, b. These are generally the rated quantities of the system which enable the converter to establish its own voltage and frequency in order to establish the grid (i.e., the converters operate in grid-forming mode). In autonomous mode, magnetization support of DFIG has to be provided with capacitor banks. Instead of the DFIG rotor side converter connecting to the battery as in Figure 2.16a, it can be connected to the DC bus of a hybrid microgrid (refer Figure 2.6); this alternative can provide an additional corridor for power exchange between the AC and DC networks in a hybrid microgrid.

On the contrary, PMSG will be a natural choice to be part of autonomous microgrid systems, as it does not require any external magnetization support, unlike DFIG. It has inherent black start capability too. In the case of AC grid operation of PMSG, the rated voltage and frequency can be established by the grid-side converter through grid-forming control. As earlier, the rotor side converter can be an uncontrolled rectifier, but if MPPT is intended then it has to be a controlled voltage source converter (VSC).

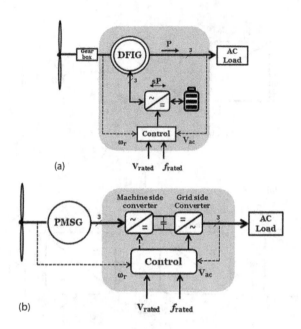

(a)

(b)

FIGURE 2.16   Wind powered AC autonomous microgrids using (a) DFIG and (b) PMSG.

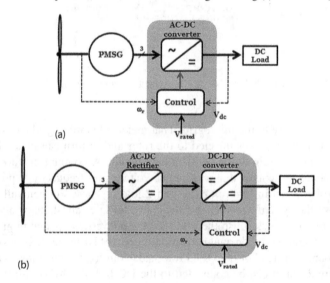

(a)

(b)

FIGURE 2.17   PMSG based DC autonomous microgrids using (a) controlled AC-DC converter and (b) controlled DC-DC converter.

Operation of PMSG in a DC grid can have several options in terms of converter topologies; a couple of the most preferred schemes are presented in Figure 2.17a and b. The system of Figure 2.17a uses an AC-DC converter to establish a DC output which has the additional responsibility to maintain high power factor at the generator side with a suitable power factor correction algorithm. The other scheme has the unique advantage of power factor correction and DC grid establishment by control of a single switch in the DC-DC converter. In terms of control complexity and power circuit components count, the latter is a better option.

2.2.2.3   The Microgrid Controllers

The controllers in microgrids address such tasks as to manage the microgrid as a dispatchable load of the main grid, to support grid frequency control, to regulate voltage and frequency within the microgrid, to maintain power balance, to enable seamless islanding, to reconnect with soft start capability, etc.

2.2.2.3.1   Primary and Tertiary Control

Frequency and voltage control is considered to be the primary control executed through the power converters of the RE sources. Primary control is generally realized through a centralized, distributed or hierarchical approach, considering the local generation and load conditions to manage power balance within the microgrid and to enable power exchange with the main grid. Further, there will be a tertiary control specifically targeting the economic and optimization aspects of the microgrid such as storage management, RE generation control and dispatch, resource optimization, forecasting RE generation and load, etc.

2.2.2.3.2   Load Control

Microgrids are appended with the load control at the end user site such that the load profile follows the available generation and storage. The non-essential loads, as declared by the consumers like water pumps, water heaters, battery chargers, etc., can be remotely turned off while maintaining the power supply to critical loads such as life-support equipment, streetlights and computer servers. Also, load control provides opportunity for the consumers to benefit from arbitrage power tariffs and to cut down their electricity bills.

2.2.2.3.3   Islanding Control

Interconnected microgrids include an islanding detection and control to facilitate isolation from and reconnection to the main grid as demanded by the operating conditions. An islanding detection unit will always monitor the presence of grid and control signals will be generated and sent to trip the static switch at the point of common coupling (PCC) to disconnect from the main grid. Often, the microgrids are islanded intentionally for tariff benefits, improved power quality and better reliability; such decisions are to be made by the tertiary control by investigating the operating conditions of the microgrid. Further, soft start reconnect is an advanced feature to control the active power ramping rate when the microgrid resumes its grid integration following a grid fault clearance. This control feature in microgrids are essential to avoid further failure of the main grid due to large number of microgrids simultaneously reconnecting after an outage and generating the rated power.

2.2.3 Operating Modes of Converters in Microgrid

Converters are the entities through which various RE sources are connected in parallel to form the microgrid, which is often connected to the main grid. Thus, the converters are to be built with the control capability to work in parallel to one another and in synchronism with the grid. When any of the RE sources trips and the main grid also goes off, then the converters are expected to continue their support to essential loads in stand-alone mode. Similarly, the local energy storage systems is also connected to the microgrid through bidirectional converters, which control the power flow in either direction. There are three modes of operation of various microgrid converters; namely, grid-feeding mode, grid-support mode, and grid-forming mode. The first two modes are operable with grid-connected condition while the third mode is required under autonomous operation.

2.2.3.1 Grid-Feeding Mode

In grid-feeding operation, the microgrid is connected to the main grid at PCC to facilitate power exchange with the latter. The microgrid in this mode can be regarded either as a source when injecting excess power into the main grid or as a load when absorbing power from the main grid. Grid-feeding converters are operated to deliver a controlled amount of active and reactive powers as commanded by the local controller. The converter in this mode will act as a current source delivering to the main grid a current of a magnitude proportional to an active power command and at a phase deduced from a reactive power command. The power converter needs here a perfect synchronization with the grid for a regulated active–reactive power delivery to the grid. As such, the converters with grid-feeding control cannot work in autonomous condition and it is necessary to have at least one synchronous generator within the microgrid to establish the grid voltage in order to provide synchronization reference for the grid-feeding type of converters. In the future inverter dominated grids, the responsibility of establishing the grid may be entrusted with one of the power converters enabled with grid-forming control; consequently, all the other grid-feeding converters will follow this to feed power to the formed grid/microgrid.

2.2.3.2 Grid-Forming Mode

When the microgrid is disconnected from the main grid due to grid fault or for scheduled maintenance and in spite of it if the local RE generators continue to operate with the loads and the energy storage systems, it is termed as islanded operation. Grid-forming mode of converter operation is obligatory when the microgrid is islanded and it is necessary to supply power to the loads.

Converters working in the grid-forming mode can be considered as voltage sources with output voltage and frequency established at the rated values, accomplished by converter controls enabled with predefined reference voltage and frequency. Grid-forming converters are designed and controlled to share power among these when working in parallel. With no power support from the main grid, the power balance in the microgrid is accomplished through (i) load shedding/control, (ii) regulation of converter output power, and (iii) charge-discharge control of storage devices.

2.2.3.3 Grid-Support Mode

Grid-support mode can be initiated during grid-connected or islanded operation of the microgrid based on a specific support service requirement. The major support services provided in the grid-connected mode by the inverters are: (i) injection of appropriate value of reactive power to regulate the grid voltage, (ii) real power curtailment or control for grid frequency regulation, (iii) low/high voltage ride-through (L/HVRT), requiring the inverter to stay connected with the microgrid for a prescribed time period following an event of voltage dip or rise, (iv) low/high frequency ride through, requiring the inverter to stay connected with the microgrid for a prescribed time period following an event of frequency dip or rise, and (v) specific power factor operation for sourcing or sinking VAR and maintaining grid voltage. The support services provided by the inverter in the islanded mode are: (i) frequency droop control for active power sharing between inverters, (ii) voltage droop control for reactive power support, and (iii) black start capability.

2.2.3.4 Seamless Transfer across Modes

There are various operational and control challenges for inverters when working across grid-feeding, forming and support modes, especially during the transition periods. For instance, when a transition from grid-feeding to grid-forming mode occurs, the power exchange at the PCC will be abruptly interrupted. If the power was flowing into the microgrid before this changeover, a deficit of power (equal to the amount of power exchanged prior to the mode change) occurs within the microgrid. In the first instance, a sharp decline in the microgrid frequency will occur which may cause some RE inverters to trip and can eventually lead to further decline of the frequency initiating cascaded disconnection of several micro sources until the entire microgrid fails. On the other hand, if the power was flowing out of the microgrid prior to the changeover, a surplus of power arises with a sharp rise in frequency within the microgrid after its transition to islanded mode.

However, a seamless transfer from grid feed to islanded mode can be accomplished through imparting a momentary power buffer by the local energy storage system during this transient period. Accordingly, microgrid control should be appended with one or more of the seamless transfer attributes. For example, when there is power deficit at the instant of islanding, then the load control should shed all the non-essential loads, simultaneously increasing the power output of the converters connected to energy storage system. Transient support of required duration can be achieved by an apt combination of heterogeneous energy storage system considering the different change-discharge characteristics. This can ensure the transient power balance of the microgrid immediately after a grid outage for the autonomous mode of operation. Conversely, when power is surplus at the instant of islanding, then the generation control has to come into operation and curtail the output power of one or more of the local generators and energy storage system. Else, the energy storage system can be made to receive the surplus power, if not fully charged. Also, the output of selected RE generators can be reduced by shifting their operating point off the maximum power point or can be disconnected to maintain the transient power balanced. Such control is fast and easy with solar PV, but difficult with WTG. This can save the microgrid from cascaded failure and can maintain continuity of power supply within the island. Once the transient period is elapsed, then the microgrid main controller can continue with its regular control actions according to the desired mode of operation and its dictated control tasks. Such controls can be added in a master-slave manner so as to facilitate seamless transfer from any mode to any other mode.

Migration from grid feed mode to grid support is relatively simple, as the microgrid is already connected to the main grid. It happens whenever a frequency or voltage drift on the main grid (beyond the permissible limits) indicates either power imbalance or inadequate reactive power management. All the RE sources within the microgrid normally deliver controlled active–reactive powers as activated through converter command signals. Therefore, the grid-support services like active power delivery or support, reactive power drain or absorption and constant power factor operation may be translated as commands for the power converters of various micro sources. The delay perceived in the mode change is solely due to the computation time at the grid operator level. Once the command is received, it doesn't require any significant time delay to extend these supports thanks to the modern-day digital controllers operating at high clock frequencies and serving as controllers for power converters.

2.3   RENEWABLE ENERGY INTEGRATION ON MICROGRIDS

Higher RE penetration can help reduce electricity generation from fossil fuels, but it poses several integration challenges on grid operators in balancing the generation with demand. The instantaneous power penetrations of several countries have already crossed 25%, with Germany and Denmark touching even 100% though occasionally. The projected RE penetration in India is 57% of the total electricity generation by 2027. Such huge levels of RE are injected into the traditional power systems, which are primarily designed for generation following the load. Now, the variability and uncertainty of generation added to the load variability introduce new challenges to the grid operators. Therefore, the electricity authorities of various countries have come up with standards for grid connection of RE sources to achieve large RE penetration without compromising on grid stability and power quality. Accordingly, several of the grid integration challenges are effectively addressed through approaches like generation and load forecasting, demand response, RE generation control and demand dispatch. Several mitigation techniques are accomplished predominantly through power electronics by expanding their control capabilities to fulfill the utility aspirations which can be realized by information and communication technology (ICT) applications.

The distributed generation involving recent microgrid technologies is expected to meet several standards in order to ensure the grid reliability, support and safety, while maintaining smooth power transfer under all operating conditions. International agencies are developing standards for grid integration of various RE sources which are to be adopted by the utility companies of various countries. Institute of Electrical and Electronics Engineers (IEEE), International Electrotechnical Commission (IEC), German Commission for Electrical, and, Electronic and Information Technologies of DIN and VDE (DKE) are developing such standards. Microgrid equipment manufacturers have to follow these standards in order to ensure proper deliverables of their products such as meeting power factor requirements, isolation, power quality, L/HVRT capability, etc.

2.3.1 Grid Synchronization Tools and Approaches

The general structure of grid-connected converter control is presented in Figure 2.18. Most of the microgrid power converter assets will be working in grid-feed mode, irrespective of the microgrid operating mode, because in all cases the grid will be formed either by a synchronous generator or by a grid-forming converter. So, the converters working in grid-feed mode will have a grid voltage form, phase and frequency tracking system to deliver a reference signal representing the fundamental quantity of the grid voltage in order to establish synchronization with the grid. Such tracking system is called a *grid synchronizer*, which is expected to possess a high level of accuracy; else, the grid-connected system may collapse due to large circulating currents causing the protection systems to trip. The grid voltage template provided by the grid synchronizer is used to form the current reference for the inverter control. Subsequently, a current controller will make the actual inverter current to follow this reference by an appropriate PWM switching control. In the case of operations in microgrids, which are dominated by power converter fed loads and sources, the synchronizer is expected to work with non-stationary and time-varying grid signals that involve harmonics, imbalances, power quality issues such as sags and swells, and transient events like surges, etc. The synchronizer is expected to provide pure sine wave information rejecting all the power quality issues in order to maintain harmonics-free power delivery. In case of sag and swell events, the inverter currents may tend to deviate from the reference values if a proper synchronization template is not available.

Various algorithms and techniques, ranging from the basic to the advanced ones, are in use to detect accurate unit sine templates representing the grid signal with excellent dynamic response, in compliance with the grid code. Synchronizers use either frequency or time domain techniques to track the phase, form and frequency of the grid signals. Such techniques include voltage zero crossing detection, Kalman filter, discrete Fourier transform, non-linear least square, adaptive notch filtering, artificial intelligence, phase locked loop (PLL), frequency locked loop and the recent multicomponent signal processing techniques (refer to http://dx.doi.

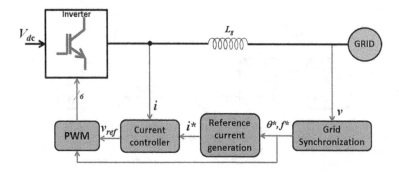

FIGURE 2.18   General structure for control of grid-connected converters.

org/10.1016/j.rser.2016.01.066). Zero crossing detection and PLL have been the most explored of these techniques; both have undergone numerous reformations in the past to suit several non-ideal grid conditions.

2.3.1.1 Voltage Zero-Crossing Detection Technique

This technique is simple and one of the earliest methods for detecting the zero cross-overs of the grid voltage signal and generating a corresponding unit sine template which can be further used by the current control loops of the grid-connected inverter as seen in Figure 2.18. This technique cannot provide a reliable synchronization signal when the grid signal has harmonics, sags and swells, frequency drifts, transients involving multiple zero crossings, etc. Voltage zero-crossing detection technique has reformed over the past years and been appended with several digital filtering techniques – namely, Kalman filter, discrete Fourier transform, non-linear least square, adaptive notch filtering, etc., to enhance its performance under various real time grid conditions.

2.3.1.2 Basic Phase Locked Loop

The PLL is the next generation synchronization method that showed better immunity to harmonics, random transient events, minor frequency deviations, etc., than the voltage zero-crossing detection; it also has showed better tracking accuracy for time varying fundamental quantities.

The basic PLL will have three components: (i) Phase detect (PD), (ii) loop filter (LF), (iii) voltage control oscillator (VCO) or phase angle generator. The PD will obtain the angle difference between the grid voltage phase (θ_{grid}) and the voltage phase detected by PLL (θ'_{grid}), and subsequently this difference is forced to become zero by LF and PI controller. This frequency difference is added to the nominal frequency value to obtain the grid frequency; integrating this frequency value results in the phase of the grid. This general scheme can be implemented in any frame of reference like stationary reference frame, synchronous reference frame, etc. Figure 2.19 gives the basic synchronous reference frame PLL (SRF-PLL) scheme.

Equation 2.1 presents the grid voltage in synchronous d-q frame of reference, while equations 2.2 and 2.3 present the respective positive sequence d and q components at the output of the PD block. The PD block takes two phase information, (i) the present PLL phase value (θ'_{grid}), and, (ii) the present grid phase value (θ_{grid}), for its dq conversion. If the grid phase is accurately tracked by dq-PLL, then the sine term

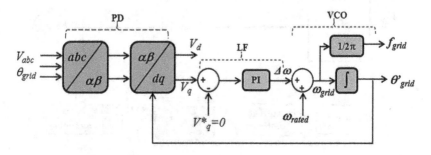

FIGURE 2.19   Basic SRF PLL scheme.

in equation 2.3 will become zero, thus the closed loop dq-PLL scheme is designed to drive V_q to become zero. Under this condition, SRF-PLL will obtain the magnitude of grid voltage as per equation 2.2. To acquire the phase of the input voltages, the q-component of the positive sequence voltage in equation 2.3 is made to track a zero reference through a PI controller and the LF as in Figure 2.19. Once the grid frequency (ω_{grid}) is obtained, then its integration will deliver the required grid phase.

$$\begin{bmatrix} v_d \\ v_q \end{bmatrix} = \frac{2}{3} \begin{bmatrix} \cos(\theta_{grid}) & \cos(\theta_{grid} - 2\pi/3) & \cos(\theta_{grid} + 2\pi/3) \\ -\sin(\theta_{grid}) & -\sin(\theta_{grid} - 2\pi/3) & \sin(\theta_{grid} + 2\pi/3) \end{bmatrix} \begin{bmatrix} v_a \\ v_b \\ v_c \end{bmatrix} \quad (2.1)$$

$$v_d = V_m \cos(\theta_{grid} - \theta'_{grid}) \quad (2.2)$$

$$v_q = V_m \sin(\theta_{grid} - \theta'_{grid}) \quad (2.3)$$

The synchronous reference frame in dq-PLL is designed to rotate only at the fundamental frequency following the positive sequence of grid voltage. Hence, if the grid signal has multiple frequencies owing to the presence of harmonics, then SRF-PLL fails to deliver an accurate synchronization template. Additionally, the tracking fails when the grid voltage is unbalanced, as a negative sequence component being appended with the existing fundamental frequency component.

2.3.1.3   Advanced PLL Structures

Owing to computational limitations, SRF-PLL fail to achieve accurate reference signals under all grid conditions. The PLL structures have continuously been improved in terms of handling varieties of adverse grid conditions, transient response and accuracy. Adverse grid conditions include presence of harmonics, sub harmonics and negative sequence components in the grid voltage. Generally, SRF-PLL set high bandwidth for the feedback loop so as to accomplish agile transient response and better performance in rejection of voltage sag/swell as well as higher order harmonics. However, accurate detection of the phase and frequency of the grid voltage with lower order harmonics will be possible by slightly reducing the bandwidth. Though reduction in bandwidth improves detection accuracy, it weakens the transient response characteristics, especially with imbalance and higher order harmonics. Also, reduced bandwidths of PLL will make them miss getting locked to the grid frequency during the transient periods and in start-up conditions. Such loss of instantaneous tracking of phase angle will eventually result in double frequency oscillations in both d- and q-axis voltages (direct axis and quadrature axis). All the advanced PLL are designed to handle one or more of the adverse grid signals simultaneously improving their steady state accuracy and transient response.

2.3.1.3.1   PQ-PLL

The issues of harmonics are better addressed through instantaneous PQ theory based PLL called the PQ-PLL or $\alpha\beta$-PLL to avail improved steady state response. In the PQ-PLL of Figure 2.20, the grid voltage V_{abc} is transformed to the stationary

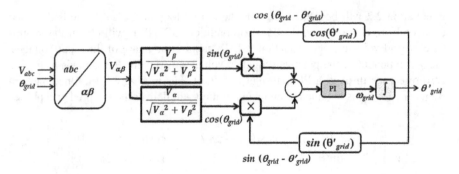

FIGURE 2.20 PQ-PLL.

reference frame using equation 2.1 with $\theta_{grid} = 0$. The phase detection of this PLL uses the trigonometric conversion $sin(\theta_{grid} - \theta'_{grid})$ deduced from $(\sin \theta_{grid} \cos \theta'_{grid} - \cos \theta_{grid} \sin \theta'_{grid})$. The PI controller compensates the phase-angle error and subsequently generates the frequency difference as the output. This frequency difference when added to the nominal frequency and integrated, yields the required phase angle θ'_{grid}.

2.3.1.3.2 DSF-PLL

The next version of PLL that could eliminate the detection errors is the Double Synchronous Frame PLL (DSF-PLL). The DSF-PLL can capture both positive and negative sequence components of an adverse grid signal. The DSF-PLL will have two SRF-PLLs, one for the positive and the other for the negative rotating frames. Various voltage vectors of the DSF are presented in Figure 2.21. The positive and negative rotating reference frames are dq^P and dq^N, each rotating at ω_{rated} and $-\omega_{rated}$, respectively; the positive and the negative frames' angular positions are θ_{grid} and $-\theta_{grid}$ respectively. Subsequently, it uses two decoupling networks as in Figure 2.22, which are designed to

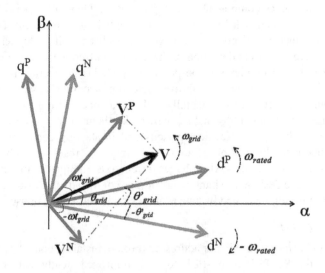

FIGURE 2.21 Vectors of double synchronous reference frame.

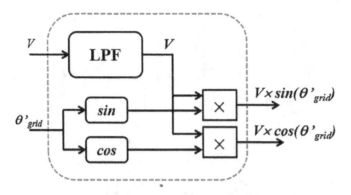

FIGURE 2.22   Decoupling network.

filter out the double frequency oscillations created in v_q by the unbalanced grid voltage in SRF-PLL of Figure 2.19. The output of the decoupling networks will provide positive and the negative sequence voltages which are presented to the two basic SRF-PLLs of DSF-PLL. The positive sequence and negative sequence values of V_d are used to extract the magnitude information as shown in Figure 2.19. The negative sequence V_q from the decoupling network is used to obtain the phase information.

2.3.1.3.3   SOGI-PLL

The further refined form of PLL is the second order generalized integrator PLL (SOGI-PLL), wherein the fundamental positive sequence component is extracted with a second order generalized integrator and presented to the basic SRF-PLL. An accurate sequence component extraction needs 90° phase shift between the stationary reference frame grid voltages V_α and V_β. This phase shift in SOGI-PLL is realized through a quadrature signal generator using a second order generalized integrator. The block diagram of the SOGI-PLL is presented in Figure 2.23a, where $V_{\alpha\beta}{}^P$ represents the positive sequence voltages in stationary reference frame, qV_α and qV_β are their respective quadrature components with the SOGI block performing a second order integration as in Figure 2.23b.

However, the quadrature generator should adapt itself to the frequency variations of the grid; else it may cause error in calculation of the sequence components, V_α and V_β. Also, it should be designed to attenuate harmonics in the grid signal. In SOGI-PLL, a combination of low pass and band pass filters is used wherein the former introduces a 90° phase shift along with harmonic filtering while the latter provides only harmonic filtering. The transfer functions of V_α to V and qV_α to V characterize a band pass filter and a low pass filter respectively as,

$$\frac{V_\alpha(s)}{V(s)} = \frac{k\omega_{grid}s}{s^2 + k\omega_{grid}s + \omega_{grid}{}^2} \; ; \; \frac{qV_\alpha(s)}{V(s)} = \frac{k\omega_{grid}{}^2}{s^2 + k\omega_{grid}s + \omega_{grid}{}^2} \qquad (2.4)$$

where k is the pass band filter gain and ω_{grid} is the grid frequency. The ω_{grid} in Figure 2.23a is made available from the PLL output as an update for the filters so that SOGI adapts to the dynamic grid conditions.

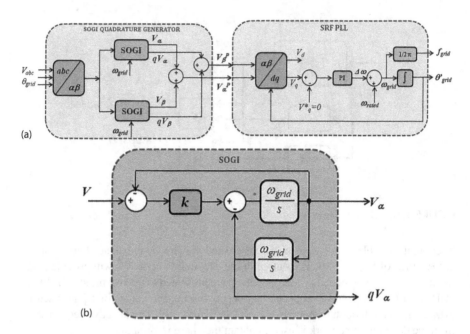

(a)

(b)

FIGURE 2.23 (a) The structure of SOGI-PLL and (b) the SOGI block.

2.3.1.4   Grid Synchronization with Advanced Signal Processing Tools

The improvements in all the advanced PLL structures attempt to lock the phase or
frequency of the grid voltage during the entire operating range. However, the modern grid operating conditions do not entail constant voltage, frequency and phase.
Especially with increased penetration of distributed generation, the main grid as well
as microgrids suffers from capricious voltage and frequency profiles and is prone to
dynamic instability. Even the reference frame transformation and the advanced PLL
structures may underperform in tracking the varying grid profiles. Multicomponent
signal processing offers better solution here.

Synchronization can be recognized as fetching a single frequency component of a
grid voltage, which may have multiple frequencies in the presence of harmonics. Signal
processing has recently seen a significant advancement in analyzing non-stationary
multicomponent signals, yet not bridged with the requirements of power electronic
control. Any signal that is comprised of a collection of many individual frequency components with a clear depiction in time-frequency plane and predefined relationships
in one or more characteristics such as frequency, magnitude, phase, etc., are called
Multi Component Signals (MCS). The grid signals of microgrids are envisaged to be
MCS, as they may contain multiple harmonics with varying combinations over time,
besides small band variations in fundamental frequency. Any MCS can be decomposed
as various frequency components in time-frequency plane using adaptive mathematical
techniques. These decomposition techniques possess the ability to recognize the relationships when moving between time domain and frequency domain. These features of
the decomposition techniques can be potentially used in the future grid synchronizer for

accurate extraction of the fundamental frequency component from grid. The decomposition techniques with such features will be potentially useful in the future grid synchronizers for accurate extraction of the fundamental grid frequency.

2.3.1.4.1 Wavelet Transforms

One of the fundamental time-frequency decomposition algorithms is wavelet transform. Wavelets are capable of fetching the fundamental information from the MCS of grid voltage as a time series and wavlet transform can perform as an effective grid synchronizer. The wavlet transform decomposes any signal into various scales of a mother wavelet in time-domain in a variable observation window and then forms the local structure in time-frequency domain. The wavlet transform is designed to fetch the fundamental frequency as a time series data in synchronization applications, so that a closed loop controller of a converter can utilize this time series data to develop the reference template just like PLL. The wavlet transform synchronizer can precisely extract the fundamental component from distorted grid signals in spite of multiple power quality issues. Yet, the non-adaptive nature of wavelets gives rise to erroneous synchronization signals during transient conditions; it demands further refinement of the signal decomposition technique.

2.3.1.4.2 Empirical Wavelet Transforms

Empirical mode decomposition (EMD) is an adaptive method that uses the information contained in the signal to decompose the MCS effectively even when it is highly non-linear and non-stationary. But as the ad hoc nature of its algorithm lacks mathematical theory, EMD cannot be used directly for synchronization application. Yet it led to the development of adaptive WT, called the empirical wavelet transforms (EWT). A set of band pass filters are included in EWT in such a way that these capture the location in the spectrum where the information is available. This helps in precisely segregating the fundamental component from the grid signals even in the presence of transient events.

2.3.1.4.3 Variational Mode Decomposition

In variational mode decomposition (VMD), the MCS is divided into time series components centered around a single frequency or a mode with small sidebands. Thus, obtaining multiple time series subcomponents of any MCS at specific frequency of interest is made possible with VMD. These sidebands equip VMD to accommodate the deviation, if any, in the fundamental frequency of the grid signal. The VMD has the potential to be one of the strong candidates to be a grid synchronizer, as it can segregate the fundamental from a non-stationary grid signal even in the presence of multiple harmonics and with time varying fundamental frequency.

2.4 POWER MANAGEMENT AND CONTROL IN MICROGRIDS

A variety of electric power generators are available on various microgrids such as diesel engines, gas micro turbines, photovoltaic cells, fuel cells, and wind turbines. The power converters through which most of these generators are tied to the network also provide control functionalities required in the microgrids. These converters work as inverters when the power flow required is from the DC side to AC side and as rectifiers for a reverse power flow. The structure of these converters and their control capabilities are very much generic irrespective of the modes of operation. In grid-connected mode, the inverters are expected to control their individual power outputs as commanded by their RE maximum power point trackers (MPPT) and maintain balance within the microgrid through P/Q control.

The salient control responsibilities of such inverters are: (a) AC voltage generation, (b) independent control of active and reactive powers, (c) synchronization to grid, (d) meeting or exceeding the harmonics standards, (e) grid forming, (f) low/high voltage ride through, (g) minor voltage/frequency disturbance ride through, (h) control under grid fault and distorted grid conditions, and (i) islanding detection and isolation. The control capabilities of the grid-connected power converters are achieved through feedback controllers with a wide range of controlled and controlling parameters.

The performance of such feedback controllers of power converters is assessed based on stability, reference tracking accuracy and dynamic response. The inverter feeding the grid in the grid-connected mode is equivalent to a current source with the control objective as to deliver a predefined amount of current for a given power reference as detailed in Section 2.2.3.1. Therefore, the controller has to be designed as a current regulator and be provided with inputs as reference current and actual current. There is a conflict in this control problem that the controlled parameters are the line currents delivered by the power converters which are AC quantities whereas the legacy control techniques like P, PI, and PID regulators handle only DC quantities. So, the control challenges are: (i) Is it possible to have DC control loops rather than AC? (ii) If AC control loops are to be used then what will be the structure of the AC regulator? (iii) Will this regulator be as effective as the time-tested DC regulators? (iv) If DC regulators only are to be used, then how can be the AC quantities accommodated in DC control loops?

This section describes the possible controller structures and their requirements to answer the aforesaid questions to work satisfactorily in microgrid environment.

2.4.1 CONTROL UNDER GRID-CONNECTED MODE

2.4.1.1 General Structure of Grid-Connected Converter Control

Voltage source converters are traditionally used in electric motor drive systems applying diverse control techniques ranging from the natural reference frame control to the synchronous reference frame control and employing conventional controllers or even artificial intelligence-based controllers. With minor tweaking, many of the control techniques developed for the motor drive systems can be made suitable for utility/grid-connected applications as well. However, when operated in the grid-connected applications, the controllers have to overlook transient conditions, different types of disturbances and parameter variations due to grid conditions varying from time to time. At the same time, these have to fulfill the grid connection standards.

The detailed control scheme of the grid-connected converter is shown in Figure 2.24, wherein a DC link is shown feeding a voltage source inverter (VSI) controlled by a multi loop controller. Irrespective of the number of preceding power

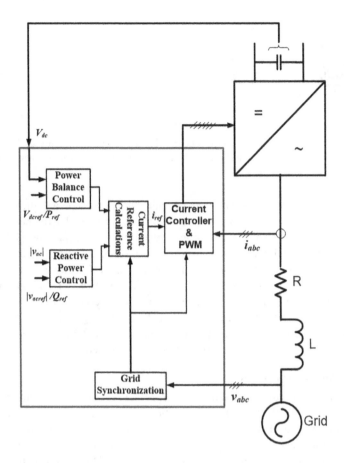

FIGURE 2.24 Detailed control scheme of grid-connected converter.

stages ahead of the DC link, the grid-side converter and its control structure will be similar to that of Figure 2.24. Most control structures share some common components and features which are discussed below.

2.4.1.2 Grid Synchronization

When an inverter has to establish and maintain an asynchronous link with the grid for a controlled power transfer, then the information about the phase, the form and the magnitude of the grid voltage need to be continuously monitored and made available to the controller. The grid voltage vector (V_{grid}) will be rotating at the rated frequency or within a specific band around the rated frequency and the inverter output voltage vector (V_{inv}) will be rotating at a different frequency prior to synchronization. That means, the relative phase angle between the grid voltage and the inverter voltage is continuously varying from 0° to 180°. At the instant of closure of the switch connecting these two active sources, the two voltage vectors should be made to rotate at the same frequency so as to avoid large circulating currents triggered due to the instantaneous voltage mismatches. Further, the phase angle (δ) between these two

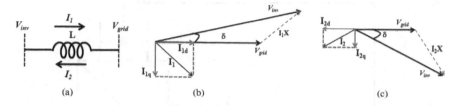

FIGURE 2.25 Grid-connected converter control and phasor diagram: (a) Equivalent circuit of inverter connected to grid, (b) Power delivery to grid, and (c) Power delivery from grid.

voltage vectors (known as power angle) decides the magnitude and direction of the current (I_1 or I_2) between the two sources, which in turn decides the power transfer as depicted in Figure 2.25. The role of grid synchronizer is imperative in generating an accurate sinusoidal current reference for the grid-connected inverters. Accurate grid information will ensure reliable synchronization and harmonic-free AC current injection. Such information is provided by a grid synchronizer through grid measurements which generate the reference template for the control schemes, whatsoever be the type, to achieve synchronization. Thus, every grid-connected converter system will have a synchronization unit as seen in Figure 2.24 which houses one of the diverse ranges of synchronization techniques discussed earlier.

2.4.1.3 Power Regulation

Power controllers regulate the active as well as the reactive power delivered by the modulation in all the three modes of operation of microgrids; namely, grid-feeding, grid-forming, and grid-support. The various control functions of grid-connected converters of single and multistage type are depicted in Figure 2.26. Multistage conversion provides a decoupling between RE generator and the grid, which relieves the grid-side converter of the burden of voltage regulation; improving the voltage regulation at the DC link will help reduction of harmonics by appropriate choice of inverter modulation index. At the same time, multistage converters need large energy storage device at its DC link; however, such large DC link can provide virtual inertia as well as fault/low voltage ride through support for microgrids.

For power regulation, the grid converter operation is to be identified first based on the mode of operation of the microgrid, and then the method to obtain the reference power can be identified. The commonly used strategies are instantaneous power balancing, maximum power point tracking (MPPT) of the RE generator, direct power control with droop characteristics, etc. Once the power reference is obtained, then either a voltage or a current control loop can be used to force the actual quantities to follow the respective references. These control methods are presented in Figure 2.27. In current control, the magnitude and the phase of the current injected will be modified as demanded by the reference power to be delivered by the converter; whereas in voltage control, the inverter voltage magnitude and its phase with respect to the grid voltage are modified as demanded by the reference power.

The next level of implementation will be the selection of the control structures like natural reference frame, stationary reference frame or synchronous reference frame.

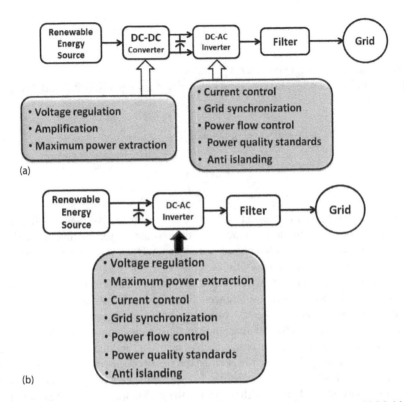

FIGURE 2.26 Control capabilities of converters in grid-connected systems: (a) Multistage converters and (b) single stage converter.

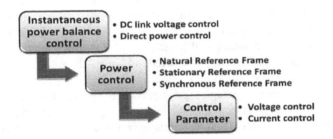

FIGURE 2.27 Control methods of grid-tied converters.

2.4.1.3.1 Active Power Control

The active and reactive power (*P* and *Q*) transfer between the inverter and the grid, which are decoupled through an inductive filter (*L*) as seen in Figure 2.25, can be expressed as,

$$P = V_{inv}I_1\cos\theta = \frac{V_{inv}^2}{\omega L}\left(\frac{V_{grid}}{V_{inv}}\sin\delta\right) \tag{2.5}$$

$$Q = V_{inv}I_1 \sin\theta = \frac{V_{inv}^2}{\omega L}\left(1 - \frac{V_{grid}}{V_{inv}}\cos\delta\right) \tag{2.6}$$

where θ is the phase angle between V_{inv} and I_1.

For small values of δ, $\sin\delta \cong \delta$ and $\cos\delta \cong 1$; then equations 2.5 and 2.6 can be reduced to,

$$P \cong \frac{V_{inv}V_{grid}}{\omega L}\delta \tag{2.7}$$

$$Q = \frac{V_{inv}(V_{inv} - V_{grid})}{\omega L} \tag{2.8}$$

It can be confirmed from equations 2.7 and 2.8, that the active power delivered by the converter depends on the angle between the voltage vectors of inverter and grid, while the reactive power delivered/absorbed depends on the algebraic difference between the voltage vectors' magnitudes.

Both active and reactive power controls are usually implemented as outer loops and provide appropriate current references to the inner control loops. The inner current loops should account the series voltage drop that occurs across the output filter when the inverter delivers the reference power. This series drop is a variable quantity and a function of the inverter current, which in turn is a function of the power. As the inverter output voltage is the vector sum of the grid voltage and the variable series drop, the control scheme should ensure that the inverter voltage vector is always higher in magnitude and phase with respect to the grid voltage vector to facilitate the power flow from DC link to the AC side.

On the other hand, in multistage RE systems, the converter on the source side is entrusted with the responsibility of extracting maximum power from the source. Typically, MPPT is provided to keep the operating point of the RE generator around MPP. However, only if the power reference of the inverter control is identified rightly at MPP, the tracking operation will be successful. One such method is instantaneous DC link power balance that is depicted in Figure 2.28; it can identify the right inverter current reference for the MPP power.

Basically, the capacitor voltage V_{dc} depends on the energy balance between the power received by the inverter at the DC link and the power delivered by it. The DC link voltage will be constant if and only if these two powers are made equal. If power

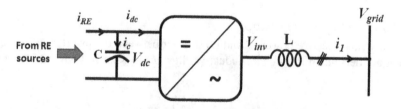

FIGURE 2.28 Concept of instantaneous DC link power balance.

received by the DC link is greater than the power delivered by it, then the extra energy will be charging the capacitor, which in turn will elevate its voltage. On the other hand, if the power delivered by the VSI is greater than the power received, then the deficit power is supplied by the DC link capacitor momentarily which results in reduction of its voltage. Thus, the DC link voltage serves as an index of power balance for the grid tied VSI, and by regulating it the maximum power extracted from the RE source can be pushed into the grid.

The DC link capacitor can be designed to operate at any desired voltage ripple and that can provide a hold-up time to facilitate the multistage grid-connected inverter system to search for the MPP. This holdup time can be the permissible MPPT delay in addition to the control and process time lag of the preceding DC-DC converter in a typical multistage grid-connected RE system. During the holdup time, the inverter keeps providing the regulated output current even during a momentary power shortage owing to a fall in irradiance in solar PV system, a fall in wind velocity in WTG system, etc.

The DC link capacitor ($C_{\text{dc-link}}$) can be designed with the assumption that the inverter has no losses; its value can be expressed as a function of the desired hold up time as,

$$C_{\text{dc-link}} = 2 \times \frac{P_{\text{rated}} \times \text{holdup time}}{(V_{\text{dc,nominal}}^2 - V_{\text{dc,min}}^2)} \qquad (2.9)$$

where $V_{\text{dc,min}}$ is the minimum voltage across $C_{\text{dc-link}}$ as chosen by the allowed ripple in the DC link voltage, typically in the range of 75%–90% of its nominal voltage, $V_{\text{dc,nominal}}$, and, P_{rated} is the rated power delivered to grid. The product of P_{rated} and *holdup time* in Equation 2.9 decides the size of $C_{\text{dc-link}}$. The transient energy support (ΔE) that the capacitor should receive or deliver to restore V_{dc} as its nominal value is then expressed as,

$$\Delta E = P_{\text{rated}} \times \text{holdup time} = \frac{(V_{\text{dc,nominal}}^2 - V_{\text{dc,min}}^2)}{2} C \qquad (2.10)$$

Generally, the volume of capacitor is proportional to its voltage rating while its energy storage capacity is proportional to the square of the voltage rating. Large holdup time results in large size of $C_{\text{dc-link}}$, at the same time, gives adequate time for MPP search; also, the converters can be designed to operate with lower switching frequency. Conversely, small holdup time though reduces the size of $C_{\text{dc-link}}$, demands higher switching frequency and superior transient response of MPPT.

2.4.1.3.2 Reactive Power Control

Inverters can be made to deliver currents at any phase angle with respect to the grid voltage by generating an inverter voltage at appropriate amplitude and phase. This feature that realizes VSI to feed power at any desired power factor helps the reactive power control. Reactive power control of RE inverters in microgrids can dynamically compensate the reactive power where it is needed the most, i.e., close to the loads and thus perform voltage control. Such distributed voltage controls significantly improve the voltage profile within the microgrid; it can further be extended as an ancillary service to the main grid which outperforms the centralized voltage control generally carried out in conventional power system.

Reactive power control is performed by the inverter according to one or more of the following requirements: (i) The reactive power support required from the converter, (ii) the power factor at which the current need to be delivered, and (iii) voltage control on the AC grid. Inverters can be made to consume or deliver reactive power based on the system needs. In microgrids often both consumption as well as delivery of reactive power will be required when operated in grid-support and grid-forming modes of operations. The control of grid tied inverter in any of the methods specified in Figure 2.27 will include a reactive power loop with a reference which can be varied in accordance with the microgrid system demand. This will ensure the required reactive power be provided by the inverter.

Reactive power control is considered as an advanced inverter function as of today. However, these advanced functions will be adopted in the near future as they can provide ride-through capability against minor voltage fluctuations, low/high voltage and grid faults which eliminates avoidable disconnects. An advanced inverter is a standard inverter that has been enabled with advanced features with no significant increase in cost. However, the extent to which an inverter can provide reactive power support/delivery depends on the percentage real power loading of the inverter.

Reactive power capacity curve of the inverter is presented in Figure 2.29. The inverter limit circle indicates the kVA rating of the RE inverter; any vector within this circle indicates the possible active and reactive power combinations without violating the capacity constraint of the inverter. In Figure 2.29a, S_1 is the kVA delivered by the inverter for an available RE active power of P_1 and the corresponding reactive power delivered is Q_1. The possible maximum reactive power delivery at this inverter loading is Q_{1max}, and a Q demand beyond Q_{1Max} will ask for a reduction in active power to P_{derate}. Figure 2.29b corresponds to another condition of RE power transfer, where the available active power is very close to the rated capacity of the inverter. It can be seen that the maximum possible reactive power support is reduced to Q_{2Max} from Q_{1Max} of Figure 2.29a. It therefore infers that when the inverter is required to provide large reactive power support, an oversized inverter is necessary to avoid real power curtailment as well as associated revenue loss.

Conversely, when the RE generator is on part load, then the active power delivered will be below the inverter kVA rating (assuming that both generator and inverter have equal power ratings). Under such operating conditions, the remaining inverter capacity can be utilized for supplying reactive power and this will contribute to improving the RE system economics. Therefore, this control feature of inverter can provide reactive power at any time of the day, regardless if the sun shines or the wind blows.

2.4.2   Voltage and Current Control in Different Reference Frames

The aim of the grid-connected inverter is to export a controlled quantity of power with respect to the available grid voltage. The active power control is established through the in-phase current component, I_{d1}, of Figure 2.25 which may be proportional to either the microgrid power demand or the maximum power available from the RE source, i.e., the controller of Figure 2.24, operates RE units either in a constant power

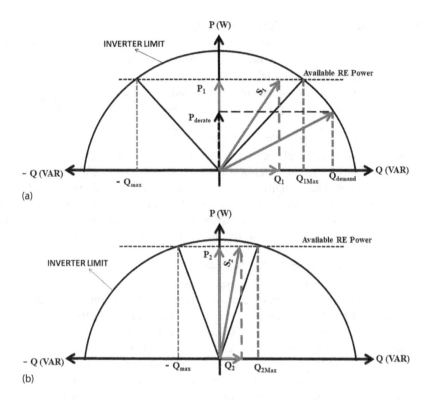

FIGURE 2.29 Reactive power capacity of RE inverter when power output is: (a) Far below the rated capacity and (b) near the rated capacity.

output or in a demand-following manner. The power commands generate a current reference for the current regulator located in the inner loop of Figure 2.24. The closed-loop controller ensures that the output current tracks the reference value with a zero steady-state error, which in turn tracks the power references. The response time of the current control loop is made relatively low, to facilitate faster current regulation. Because of the presence of outer voltage/power loop, the current limit is accomplished prior to the current control loop which eases the current regulation.

The grid-side inverter output power quality can be affected by the harmonics of the microgrid network. However, an improved quality can be targeted by proper design of the controllers as well as by appropriate filter choice. The desired characteristics of the current control includes, (i) reference tracking accuracy (ii) high bandwidth (iii) fast dynamic response, and (iv) low harmonic distortion with the capability to damp the filter resonance. These characteristics are accomplished through different control techniques for various microgrid models with diverse RE sources. Different control techniques require the system parameters in their control loops in various forms like DC, 2-phase AC, 3-phase AC, etc., but the power systems normally have 3-phase AC quantities. However, there are standard mathematical transforms used to convert the 3-phase real-time signals as required by the controller types. These transforms are revisited in the forthcoming sections.

2.4.2.1   Reference Frame Transformations

Control schemes are implemented in different reference frames; namely, *abc* or natural reference frame, stationary reference frame and synchronous reference frame. Simple control structures can be obtained through these reference frames. Ease of transformation from one reference frame to another through concrete mathematical relationships extends flexibility to carryout modeling and control in any one or more convenient frames.

2.4.2.1.1   Natural or abc Reference Frame

This is the general three phase system and does not involve any transformation. The control structure in this reference frame demands one controller each for line current regulation. The dynamic response of controllers is good in this reference frame as there are no time consuming and laborious transformations involved. The commonly used controllers for current regulation in this reference frame are hysteresis controller, proportional-resonant controller, dead-beat controllers, etc.

2.4.2.1.2   Stationary or αβ Reference Frame

This reference frame transforms three-phase *abc* signals into a two phase orthogonal system in stationary reference frame called α-β frame where both α and β axes are locked in position by Clark's transformation. In grid-tied inverter control, the three phase control variables like the line currents, grid voltages, etc., can be transformed into two phase time varying sinusoidal quantities. Working with reduced number of variables than the actual helps in simplifying the corresponding control structure, i.e., one control loop is reduced compared to the control in *abc* reference frame. The vector i_{abc} in Figure 2.30 is the space vector corresponding to the current of *abc* frame. Its α, β components are sinusoidal, but phase shifted by 90° as shown as vectors i_α and i_β.

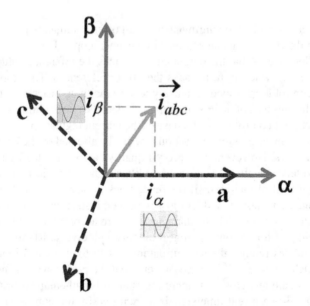

FIGURE 2.30   Stationary reference frame current vector with its α, β components.

The transformation across a balanced abc to $\alpha\beta$ reference frame currents (i_{abc} and $i_{\alpha\beta}$) is governed by the matrix ($T_{\alpha\beta}$) and the transformation is expressed as,

$$i_{\alpha\beta} = \left[T_{\alpha\beta} \right] i_{abc} \tag{2.11}$$

$$\begin{bmatrix} i_\alpha \\ i_\beta \end{bmatrix} = \frac{2}{3} \begin{bmatrix} \cos 0° & -\cos 60° & -\cos 60° \\ 0 & \sin 60° & -\sin 60° \end{bmatrix} \begin{bmatrix} i_a \\ i_b \\ i_c \end{bmatrix} = \frac{2}{3} \begin{bmatrix} 1 & -\dfrac{1}{2} & -\dfrac{1}{2} \\ 0 & \dfrac{\sqrt{3}}{2} & -\dfrac{\sqrt{3}}{2} \end{bmatrix} \begin{bmatrix} i_a \\ i_b \\ i_c \end{bmatrix} \tag{2.12}$$

The scaling factor 2/3 is considered for amplitude invariant transformations. Further, the simplified inverse transform for converting the quantities back to abc is given as,

$$\begin{bmatrix} i_a \\ i_b \\ i_c \end{bmatrix} = \frac{3}{2} \begin{bmatrix} \frac{2}{3} & 0 \\ -\frac{1}{3} & \frac{1}{\sqrt{3}} \\ -\frac{1}{3} & -\frac{1}{\sqrt{3}} \end{bmatrix} \begin{bmatrix} i_\alpha \\ i_\beta \end{bmatrix} \tag{2.13}$$

Equations 2.12 and 2.13 can be used with either line or phase quantities. Further, the same transformations can be applied on voltages as well.

2.4.2.1.3 Synchronous or dq Reference Frame

The three phase signals represented in the stationary reference frame are time varying quantities as the frame of reference is stationary. But some of the current controllers in grid connected VSI work with DC control loops, wherein they need equivalent DC quantities representing the AC currents. If the transformed quantities need to be time invariant or DC, then the observer or the frame of reference should be rotating along with the rotating space vector representing the three-phase quantity. Such a frame of reference is called the dq or synchronous reference frame (SRF), wherein the transformation takes place from orthogonal stationary system to orthogonal rotating reference system using Park's transformation. With the transformed DC quantities, the control loop can use the legacy 1st and 2nd order controllers like PI, PID, etc. The vector i_{abc} in Figure 2.31 is the current space vector corresponding to abc frame, and its equivalent d, q components with reference to the dq reference frame rotating at a frequency ω_{grid} will be DC as shown. Importantly, the reference frame can be made to rotate at any frequency as demanded by the application. It will rotate at grid frequency in grid-tied inverter applications, but in harmonic elimination control schemes it can be rotating at the respective harmonic frequencies.

Considering the q-axis leading the d-axis by 90° as shown in Figure 2.31, the currents transformation from abc to $dq0$ reference frame (i_{abc} to i_{dq}) is governed by the matrix (T_{dq}) and the transformation is expressed as,

$$i_{dq} = \left[T_{dq} \right] i_{abc} \tag{2.14}$$

$$\begin{bmatrix} i_d \\ i_q \\ i_0 \end{bmatrix} = \frac{2}{3} \begin{bmatrix} \cos\theta & \cos\left(\theta - \frac{2\pi}{3}\right) & \cos\left(\theta + \frac{2\pi}{3}\right) \\ -\sin\theta & -\sin\left(\theta - \frac{2\pi}{3}\right) & -\sin\left(\theta + \frac{2\pi}{3}\right) \\ \frac{1}{2} & \frac{1}{2} & \frac{1}{2} \end{bmatrix} \begin{bmatrix} i_a \\ i_b \\ i_c \end{bmatrix} \tag{2.15}$$

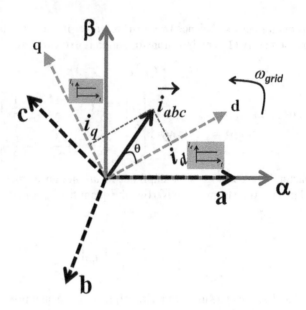

FIGURE 2.31 SRF current vector with its *d, q* components.

where θ is the reference phase angle shown in Figure 2.31. The value θ is obtained from grid synchronizers like PLL in grid-tied inverters. The inverse transformation of *dq*0 to *abc* is,

$$\begin{bmatrix} i_a \\ i_b \\ i_c \end{bmatrix} = \begin{bmatrix} \cos\theta & -\sin\theta & 1 \\ \cos(\theta - 2\pi/3) & -\sin(\theta - 2\pi/3) & 1 \\ \cos(\theta + 2\pi/3) & -\sin(\theta + 2\pi/3) & 1 \end{bmatrix} \begin{bmatrix} i_d \\ i_q \\ i_0 \end{bmatrix} \qquad (2.16)$$

The zero sequence components in equations 2.15 and 2.16 can be ignored when considering three phase three wire systems as practiced in stationary reference frame transformations.

2.4.3 CURRENT CONTROL IN GRID-CONNECTED VSI

The control loop is actuated soon after the current reference is obtained from the outer voltage/power control loop. Current control techniques commonly adopted in different modes of operation of inverters in microgrids include:

- Linear control technique
  - Proportional-Integral (PI) control
  - Proportional + Resonant (PR) control
- Predictive controls
  - Hysteresis control
  - Trajectory based control
  - Model Predictive control

- Linear quadratic control
- Sliding mode control
  - Voltage control
  - Current control
- Intelligent controls
  - Neural network
  - Fuzzy logic control

2.4.3.1 Linear Control Techniques

Linear controllers operate with either DC or AC control loops but always use conventional voltage pulse width modulators (PWM). A perfect decoupling exists between the reference current tracking control and its compensation, and also with the inverter voltage PWM. The switching frequency is constant because of PWM, and this in turn will result in an unambiguous frequency spectrum of output voltage and current.

Various linear controllers like proportional integral (PI), proportional integral derivative (PID) and proportional resonant (PR) controllers are designed with appropriate gains to accomplish reference tracking ability with very low steady state errors under all operating conditions. The controller gains are tuned according to the requirements of the grid-tied inverter and the specifications at the point of common coupling.

Because they are able to provide considerable system stability with ample phase margin, linear controllers have emerged as natural choice for voltage source inverter-control in spite of slower response and higher steady state error rates compared to modern day controllers.

With reference to Figure 2.24, the regulated output of the power controllers is received as input to the inner current controllers and is expected to produce the necessary voltage and frequency references for the PWM block which follows. The inverter output voltage is the vector sum of the AC grid voltage and the voltage drop in the series impedance. Therefore, the current controller algorithm should estimate the converter output voltage for a given power reference accounting the filter drop and simultaneously maintaining the power delivered following the reference. Some of the time-tested linear controllers of different reference frames are *hysteresis controllers* in natural reference frame PR controllers in alpha-beta and PI controllers in SRF.

2.4.3.2 Modeling of Hysteresis Current Controllers in Natural Reference Frame

In this type of control, the quantities to be controlled are maintained in their natural reference frame itself. The controlled quantities here are the phase currents i_a, i_b and i_c; these are kept in the same form and the comparison and control happen in the same frame. So, three controllers are required for controlling three phase currents, and the neutral point of the output transformer is to be connected to the inverter's ground point so as to control the phase current independently. The outer loop can have simple PI controllers as both the controlling quantities, the active–reactive powers, are mere DC quantities.

The circuit diagram of a three phase grid-connected VSI is depicted in Figure 2.32, where, V_{dc} is the DC input voltage to the VSI, u_a, u_b and u_c are the three pole voltages

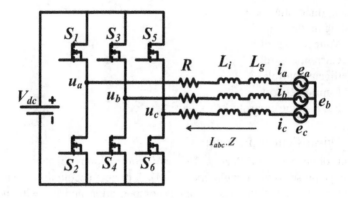

FIGURE 2.32   Three phase grid-connected VSI.

of VSI, e_a, e_b and e_c are the grid voltages, L_i is the filter inductance, R is the resistance between the inverter and the grid, and, L_g is the line inductance. The currents injected into the grid in each phase are shown as i_a, i_b and i_c.

A hysteresis current controller is adopted here in which the current delivered by the inverter is made to remain within a tolerance band around its reference value. This controller can be modeled by the load model equations of a grid-connected VSI expressed as,

$$v_{abc} = L\frac{di_{abc}}{dt} + Ri_{abc} + e_{abc} + v_{nN}\,[I] \tag{2.17}$$

where, $L = L_i + L_g$, I = identity matrix and v_{nN} = voltage at load neutral with respect to negative DC bus = $\frac{1}{3}(v_a + v_b + v_c)$.

The reference voltage is,

$$v_{abc}^* = L\frac{di_{abc}^*}{dt} + Ri_{abc}^* + e_{abc}. \tag{2.18}$$

Then the instantaneous current error is,

$$v_{abc} - v_{abc}^* = L\frac{d(i_{abc} - i_{abc}^*)}{dt} + R(i_{abc} - i_{abc}^*) + v_{nN}\,[I]. \tag{2.19}$$

If,

$$i_{abc} - i_{abc}^* = \Delta i_{abc} \tag{2.20}$$

then, equation (2.19) can be rewritten as

$$v_{abc} - v_{abc}^* = L\frac{d\Delta i_{abc}}{dt} + R\Delta i_{abc} + v_{nN}[I] \tag{2.21}$$

To remove the dependency of phase current on v_{nN} and to control these two independently, a decoupling term $\Delta i''$ is introduced such that,

$$v_{nN}\,[I] = -\left(L\frac{d\Delta i''_{abc}}{dt} + R\,\Delta i''_{abc} \right) \tag{2.22}$$

This modification will perform the control on the decoupled current error,
$\Delta i' = (\Delta i - \Delta i'')$, rather than on the total error, Δi. The final hysteresis control equations to be implemented in the inverter inner control loop are obtained using the
equations 2.17–2.22 as,

$$\left(L\frac{d\,\Delta i'_{abc}}{dt} + R\,\Delta i'_{abc} \right) = v_{abc} - v^*_{abc} \tag{2.23}$$

The current control equation (2.23) shows that all variables in it are independent of
v_{nN}, which enables independent control of each phase current. The reference current,
$i_{ref} = I_m \sin\omega t$, with the hysteresis current for the tolerance band, i_H is calculated as
$i_H = (i_{ref} \pm H)$, where H is the width of the band in A. The schematic diagram of this
control implementation is given in Figure 2.33.

2.4.3.3 Modeling of SRF Current Controllers

The single phase equivalent circuit of the three phase grid-connected VSI of
Figure 2.32 is presented in Figure 2.34.

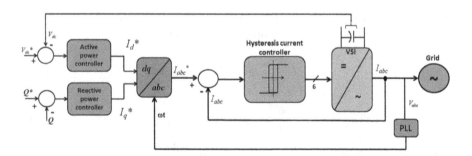

FIGURE 2.33 Grid-connected VSI control in natural reference frame with hysteresis
controller.

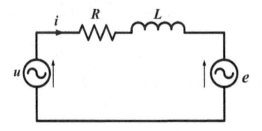

FIGURE 2.34 Single phase equivalent circuit of inverter feeding the grid.

The differential equations of the system of Figure 2.34 for the shown current/power flow direction is expressed as,

$$
\begin{bmatrix} u_a \\ u_b \\ u_c \end{bmatrix} = R \begin{bmatrix} i_a \\ i_b \\ i_c \end{bmatrix} + L \frac{d}{dt} \begin{bmatrix} i_a \\ i_b \\ i_c \end{bmatrix} + \begin{bmatrix} e_a \\ e_b \\ e_c \end{bmatrix} \tag{2.24}
$$

Equation 2.24 can be rewritten as,

$$
L \frac{di_{abc}}{dt} + Ri_{abc} = u_{abc} - e_{abc} \tag{2.25}
$$

So,

$$
L \frac{di_{abc}}{dt} + Ri_{abc} = \Delta v_{abc} \tag{2.26}
$$

where $\Delta v_{abc} = u_{abc} - e_{abc}$. The abc to dq transformation of the line currents in matrix form is given as,

$$
\begin{bmatrix} i_d \\ i_q \end{bmatrix} = \frac{2}{3} \begin{bmatrix} \cos \omega_{grid}t & \cos(\omega_{grid}t - 120°) & \cos(\omega_{grid}t + 120°) \\ -\sin \omega_{grid}t & -\sin(\omega_{grid}t - 120°) & -\sin(\omega_{grid}t + 120°) \end{bmatrix} \begin{bmatrix} i_a \\ i_b \\ i_c \end{bmatrix} \tag{2.27}
$$

where $\omega_{grid} = 2\pi f$ and f is the grid frequency. Using Equation 2.27 and its differential, equation 2.26 can be transformed into dq reference frame as,

$$
L \frac{di_d}{dt} = \Delta V_d - Ri_d + \omega Li_q \tag{2.28}
$$

$$
L \frac{di_q}{dt} = \Delta V_q - Ri_q - \omega Li_d. \tag{2.29}
$$

Substituting $\Delta V_d = u_d - e_d$, and $\Delta V_q = u_q - e_q$, the d-axis and q-axis inverter output voltages to be established to deliver the d- and q-axis currents, i_d and i_q, to the grid against the voltages, e_d and e_q are

$$
u_d = e_d + Ri_d + L \frac{di_d}{dt} - \omega Li_q \tag{2.30}
$$

$$
u_q = e_q + Ri_q + L \frac{di_q}{dt} + \omega Li_d \tag{2.31}
$$

By taking Laplace transform of equations 2.30 and 2.31, the complex transfer function $G(s)$ is derived as,

$$
G(s) = \frac{1}{(s + j\omega)L + R} \tag{2.32}
$$

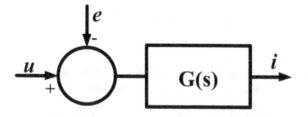

FIGURE 2.35 Block diagram of the current controller.

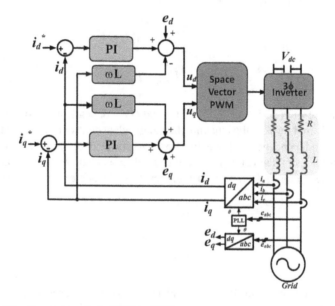

FIGURE 2.36 Current control of VSI in SRF.

With the grid voltage modeled as load disturbance, e, and the inverter pole voltage modeled as a control voltage, u, the required output current vector, I, can be obtained as shown in Figure 2.35. The power electronic circuit which supplies the voltage, u, is a VSI that produces the desired voltages with no time delay and with low harmonic distortion.

Thus the reference required for the inverter control is established using equations 2.30 and 2.31. Based on these equations, the general block diagram of current control of grid-connected VSI in SRF can be formulated as in Figure 2.36, where i_d^* and i_q^* are d and q axes reference currents, respectively. These references are obtained from the outer loop which will be similar to that in Figure 2.33. A PLL gives the necessary phase information of the grid voltage for the abc-dq transformation blocks.

If the voltage drops due to the line impedance is dynamically compensated for all power values using a PI controller, the equations for u_d and u_q can be given as,

$$u_d = e_d - \omega L i_q + \left(K_p + \frac{K_i}{s} \right)\left(i_d^* - i_d \right) \qquad (2.33)$$

$$u_q = e_q + \omega L i_d + \left(K_p + \frac{K_i}{s} \right)\left(i_q^* - i_q \right)$$

(2.34)

Thus, the inner current control loop will establish the inverter currents as commanded by the reference currents from the output power loops.

2.4.3.3.1 Decoupling of Active–Reactive Power Controllers

The active power P and the reactive power Q delivered by grid-connected VSI in SRF or dq quantities are given as,

$$P = \frac{3}{2}\left(u_d i_d + u_q i_q \right)$$

(2.35)

$$Q = \frac{3}{2}\left(u_d i_q - u_q i_d \right)$$

(2.36)

It can be seen from (2.35) and (2.36) that P and Q are dependent on both d- and q-axis currents delivered by VSI, thus making the independent control of P and Q impossible. The loss of independency in the active–reactive power control caused by such cross coupling is undesirable in any grid-tied inverter. If u_q is made zero by aligning the grid voltage space-vector to the d-axis, then the resulting decoupled power equations are,

$$P = \frac{3}{2}u_d i_d$$

(2.37)

$$Q = \frac{3}{2}u_d i_q$$

(2.38)

Now, P and Q can be controlled independently by i_d and i_q, respectively in two independent current control loops. This condition is achieved when voltage at the synchronizer's coupling point is taken as the reference for dq frame transformations. In spite of decoupling the power equations, the inverter voltage equations of 2.30 and 2.31 exhibit a cross coupling between the two due to the presence of the complex inductance drops represented by the factors, $\pm \omega L i_{dq}$. Here, the d-axis control voltage, u_d, depends not only on i_d but also on i_q and vice versa; in other words, if i_d is varied for whatsoever reason, it affects u_d as well as u_q and vice versa. It is equivalent to two first order systems interacting with each other and resulting in cross coupling. Complete independence in P and Q control is still not achieved owing to this cross coupling.

The mitigation of cross coupling is possible by cancelling the complex inductance drop by moving the pole of the plant $G(s)$ from $-\left(\frac{R}{L} + j\omega \right)$ to $-\left(\frac{R}{L} \right)$ in SRF and by adding a real zero by the compensator. This is achieved by selecting a control voltage, u_{dq}, as,

$$u_{dq} = u_{dq}^* + j\omega L i_{dq}$$

(2.39)

By substituting u_{dq} (from equation 2.39) in equation 2.30 and 2.31, a feed forward decoupling is introduced and the decoupled system control equation is obtained as,

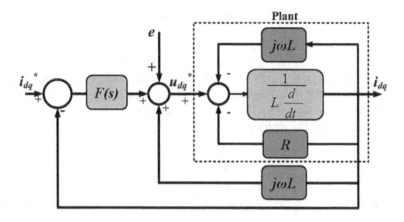

FIGURE 2.37 Current control with an inner decoupling loop.

$$L\frac{di_{dq}}{dt} = u_{dq}{}^* - Ri_{dq} - e_{dq} \qquad (2.40)$$

It is evident from equation 2.40 that there is no cross coupling as there is no complex valued coefficients. Equation 2.39 is modeled as an inner feedback loop and a current regulator having output as $u_{dq}{}^*$ is designed as an outer loop for the decoupled system.

The block diagram of the decoupled control system with the new control voltage of equation 2.39 is given in Figure 2.37, where i_{dq} is the reference vector expressed as,

$$i_{dq} = i_d^* + ji_q^* \qquad (2.41)$$

The transfer function of the decoupled system between $u_{dq}{}^*$ and i_{dq} is expressed as,

$$G'(s) = \frac{1}{sL + R} \qquad (2.42)$$

This first order complex valued system with no interacting terms can be regulated using PI controller with transfer function as $F(s) = k_p + k_i/s$. The control is not susceptible to the changes in the impedance, ωL, between the inverter and the point of common coupling (PCC), because of the feed forward decoupling term introduced in the control loops. This feed forward decoupling term removes the cross coupling between the P and Q control loops resulting in completely independent P and Q controls.

2.4.3.3.2 Tuning of the PI Controllers for the Grid-Connected VSI

The constants, K_p and K_i, decide the transfer function as well as the location of poles and zeros of the PI compensators of equations 2.33 and 2.34. Various tuning methods are available in the literature ranging from trial and error, Ziegler–Nichols, Tyreus Luyben, soft computing based tuning, loop shaping, etc. The desirable PI controller gains are obtained through the *loop shaping* method to suit the grid-connected converter system requirements and presented in this section as an example for PI tuning. Loop shaping method demands a desirable rise time to be assigned for the closed loop system based on its constituent components. For example, a rise time of 1 ms can be considered acceptable while working with Transistor/IGBT inverters, and can be further reduced when working with MOSFET devices. $G'(s)$ of equation 2.42 being a complex first order system, the resulting closed loop transfer function is intended to be obtained as first order system with closed loop bandwidth of α as,

$$G_c(s) = \frac{\alpha}{s+\alpha} \tag{2.43}$$

The standard relationship between the rise time t_r and α for a first order system is,

$$\alpha t_r = \ln 9 \tag{2.44}$$

The closed loop transfer function is obtained using the compensator transfer function, $F(s)$, and the plant transfer function, $G'(s)$, of equation 2.42. The closed loop transfer function for a negative feedback system is obtained as,

$$G_c(s) = \frac{F(s)G'(s)}{1+F(s)G'(s)} \tag{2.45}$$

So, if $F(s)G'(s)$ is selected as,

$$F(s)G'(s) = \frac{\alpha}{s} \tag{2.46}$$

Then,

$$G_c(s) = \frac{F(s)G'(s)}{1+F(s)G'(s)} = \frac{\alpha/s}{1+\alpha/s} = \frac{\alpha}{s+\alpha} \tag{2.47}$$

Equation 2.47 is found to be equal to equation 2.43, meaning that the desired closed loop response is achieved. Further, equation 2.46 yields,

$$F(s) = \frac{\alpha}{s}\left(G'(s)\right)^{-1} = \frac{\alpha}{s}\left(sL+R\right) = \alpha L + \frac{\alpha R}{s} \tag{2.48}$$

Equation 2.48 represents the transfer function of the PI controller and so the K_p and K_i values can be obtained as,

$$K_p = \alpha L$$
$$K_i = \alpha R \tag{2.49}$$

Thus, the controller parameters are expressed as the parameters of the plant transfer function, L and R, and within the required closed loop bandwidth. Such tuning of PI controller avoids the trial and error method and gives better stability for the entire bandwidth, especially for systems with PWM switching converters.

2.4.3.3.3 Performance Evaluation of Grid-Tied VSI with SRF Current Controllers

The grid-tied VSI with SRF current controllers is simulated with the specifications of Table 2.1 and the system performance like dynamic response, steady state response, frequency tracking, harmonic content, active power delivery, and reactive power consumption has been studied in this section.

- *Steady state performance*: Figures 2.38 and 2.39 shows the current delivered to the grid and the voltage at PCC for a power reference of 275 W till 0.6 s and then subjected to a step change to 1450 W. That the rise time of the step change response is less than 2.5 ms and shows the ability of the controller to track step changes.
- *Harmonic analysis of the injected current*: Figure 2.40 shows the FFT analysis of the grid current as a frequency spectrum. THD is only 0.17%. All the components are less than 3% which satisfies all grid codes. The lowest order harmonic appears at the switching frequency of 4 kHz and the subsequent harmonics are at its multiples.

  Figure 2.41 shows the active and reactive powers delivered with step changes in the references and also the current delivered. Since power is fed at the unity power factor into the grid, the reactive power component is nearly zero for $Q_{ref} = 0$. During the period from $t = 0.6$–1.2 s, the P_{ref} is given a step change from 275 to 1450 W and it is observed in Figure 2.41 that the value of power injected increases instantaneously. But the reactive power is unaffected by the change in active power. This shows that active and reactive components of power could be independently controlled in the case of SRF-PI current control.

TABLE 2.1
Grid-Tied VSI Specifications

Rated power	2 kW
Output line voltage	400 V
Output current	1.44 A
DC link voltage	680 V
Converter	3 leg IGBT inverter
Output filter	$L = 11$ mH, $r_L = 0.23\ \Omega$
Grid frequency	50 Hz
K_p and K_i	24 and 505
Inverter switching frequency	4 kHz

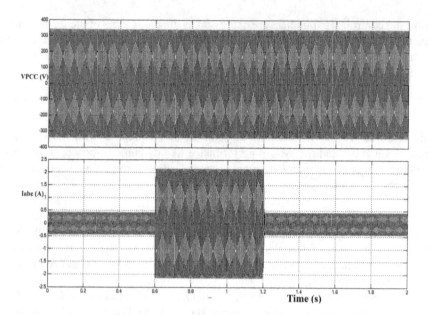

FIGURE 2.38 Steady state waveforms of current injected to the grid and voltage at PCC.

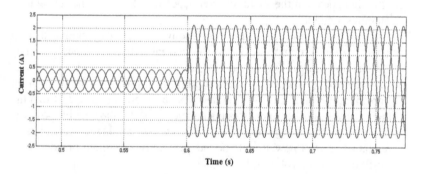

FIGURE 2.39 Steady state current zoomed.

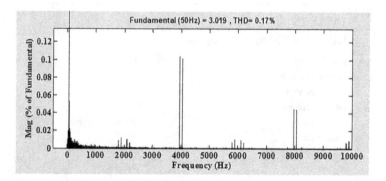

FIGURE 2.40 FFT analysis of grid current.

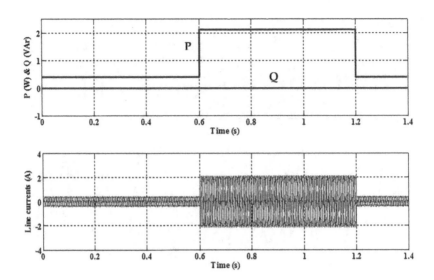

FIGURE 2.41 Active power, reactive power, and line currents.

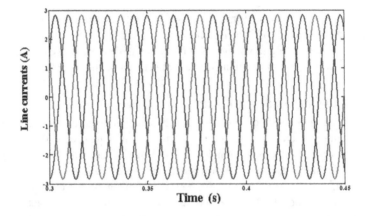

FIGURE 2.42 Line current with step change in grid frequency.

- *Response to step change in grid frequency*: The simulated grid-connected system is tested for step change in the grid frequency from 50 to 52 Hz introduced from 0.4 to 0.6 s; Figure 2.42 shows the corresponding variation in grid current. The frequency change in current waveform is visible in the figure. The current change is instantaneous and without losing stability. The magnitude of injected current remains the same in spite of frequency change. This shows that the controller is immune to frequency changes. Such step change in frequency is not practical, yet this test has been done to demonstrate the precision of tracking.
- *Response to step change in grid voltage*: The system is then tested for variation in grid voltage as this is a common scenario expected in microgrids.

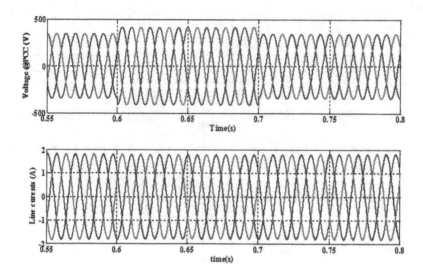

FIGURE 2.43 Inverter voltage and current response to step change in grid voltage.

A swell in grid voltage from 220 to 240 V is introduced from 0.6 to 0.7 s as depicted in Figure 2.43. The line currents delivered show no variation in spite of the variations in grid voltage. A corresponding increase in the inverter voltage is observed, in order to inject the same current at the higher grid voltage.

2.4.3.4 Modeling of PR Controllers

PR controllers can be used in grid-connected applications with control variables converted to stationary reference frame. The computation sequence of PR controller is not complex because there is no transformation from the stationary frame to synchronous frame. The active and reactive powers can be expressed in $\alpha\beta$ reference frame as $P = v_\alpha i_\alpha + v_\beta i_\beta$ and $Q = -v_\beta i_\alpha + v_\alpha i_\beta$. So, power control can be achieved by controlling the $\alpha\beta$ current components. The three phase currents obtained as feedback signals are first converted to two phase $\alpha\beta$ components and then compared with $\alpha\beta$ reference values. The resultant error signal is sinusoidal and not DC as in the case of SRF-PI controllers. A PR controller has the capability to track AC sinusoidal quantities with zero steady state error and without any phase delay. It can be inferred from the transfer function of PR controller in equation 2.50 that it becomes a simple PI controller when $\omega = 0$. Thus, PR controller can be viewed as a generalized 2nd order AC integrator tuned to the grid frequency, ω. A PR controller can therefore handle AC quantities directly without any DC transformation. But, a PR controller is very sensitive to the grid frequency fluctuations, as it introduces infinite gain only at the tuned grid frequency. If the grid frequency drifts outside the tuned band of the controller, the system may fail to track the reference. It is possible to maintain the reference tracking with wide tuned bands, but at the cost of increased steady state error. However, PR controllers have a great advantage in comparison with other current regulators for grid-connected applications that harmonic compensation can be done without affecting the fundamental reference tracking control.

The transfer function of the PR controller is,

$$H_{PR}(s) = K_P + \frac{K_I s}{s^2 + \omega^2} \qquad (2.50)$$

where K_P is the proportional gain, K_I is integral gain and ω is the grid frequency in rad/s. The basic concept of PR controller is to introduce an infinite gain at a selected resonant frequency for eliminating the steady state error at that frequency. This is conceptually similar to an integrator whose infinite DC gain forces the DC steady-state error to become zero in PI controllers. The resonant portion of the PR controller can therefore be viewed as a generalized AC integrator. The PR transfer function can be expanded as,

$$H_{PR}(s) = K_P + \frac{1}{2}\left(\frac{K_I}{s + j\omega} + \frac{K_I}{s - j\omega} \right) \qquad (2.51)$$

In equation 2.51, $(s + jw)$ is the positive sequence integrator and $(s - jw)$ is the negative sequence integrator. The block diagram of PR controller is depicted in Figure 2.44, in which the input, $u(s)$, will be an error signal and the output, $y(s)$, will be an actuation signal.

The gain of the PR controller at resonant frequency can be varied by varying K_I, therefore K_I has to be very high for better dynamic response. The value of K_P adjusts the bandwidth of the controller while simultaneously deciding the stability of the plant. A proper selection of K_P can adjust the gain outside the tuned frequency and thus a considerable gain can be maintained even with minor grid frequency deviations. Figures 2.45 and 2.46 show the frequency responses of PR controller for various combinations of K_P and K_I values. Usually, K_I is much higher than $100\, K_P$.

The ideal PR controller gives an infinite gain at resonant frequency. The gain is brought down to usable range by introducing a damping factor δ, which also increases the bandwidth of the controller. The frequency response of the controller with the damping factor is presented in Figure 2.47, wherein the gain at resonance is found to be reduced from its undamped value. Thus, the damped PR controller is designed to operate on partially loaded (below the rated power) condition due to a huge loss in the damping resistor. The transfer function of the damped PR controller can be defined as,

$$H_{PR}(s) = K_P + \frac{sK_I}{s^2 + 2\delta\omega s + \omega^2} \qquad (2.52)$$

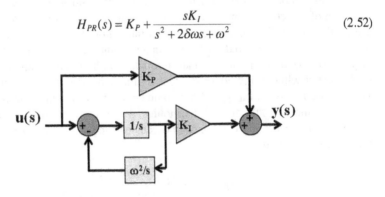

FIGURE 2.44 PR controller.

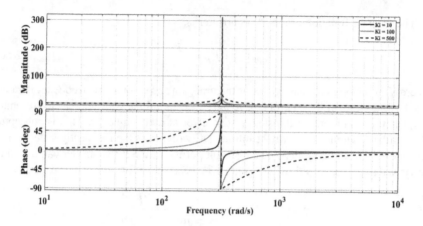

FIGURE 2.45   Frequency response of PR controller for different values of K_I with $K_P = 1$.

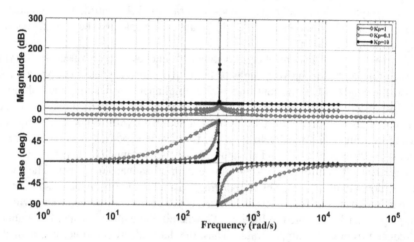

FIGURE 2.46   Frequency response of PR controller for different values of K_P with $K_I = 100$.

Another important characteristic feature of PR controller is the possibility of inclusion of selective harmonic compensation in the same control structure. This is achieved by cascading several generalized integrators tuned to resonate at the harmonic frequencies required to be eliminated. Since the PR controller acts on a very narrow band around its resonant frequency ω, harmonic compensation can be implemented without any adverse effect on the behavior of the current controller. The transfer function of a typical harmonic compensator could be designed to compensate for the 3rd, 5th, and 7th harmonics, as these are the most prominent ones in the current spectrum. The controller transfer function for compensation of any harmonic order "h" is,

$$H_{PR}(s) = K_P + \sum_{h=1,5,7,...} \frac{s K_I}{s^2 + (h\omega)^2} \qquad (2.53)$$

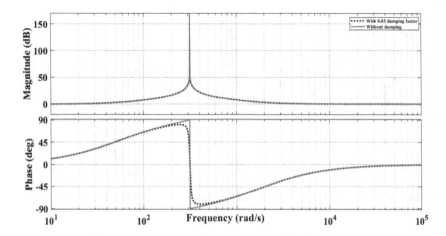

FIGURE 2.47   Frequency responses of damped and undamped PR controller.

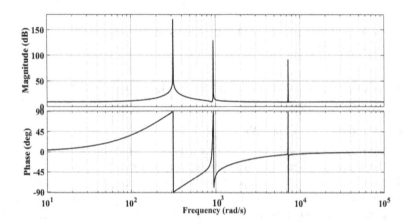

FIGURE 2.48   Frequency response of PR controller with harmonic compensation.

The frequency response of the harmonic compensated PR controller is presented in Figure 2.48, wherein it is tuned to eliminate three harmonic frequencies.

The complete control implementation of the PR controller for the grid-tied VSI with outer loops of power control and harmonic compensation is depicted in Figure 2.49.

2.4.3.5   Digital Implementation of PR Controller

The real time implementation of PR controller in a typical digital platform necessitates discretization of the transfer function. It is possible through bilinear transformation to transform the quantities into z domain by substituting $s = \frac{2}{T}\frac{(1-z^{-1})}{(1+z^{-1})}$ in equation 2.52 and the discrete transfer function of the PR controller will be,

$$G_{PR}(z) = \frac{n_0 + n_1 z^{-1} + n_2 z^{-2}}{1 + d_1 z^{-1} + d_2 z^{-2}} \tag{2.54}$$

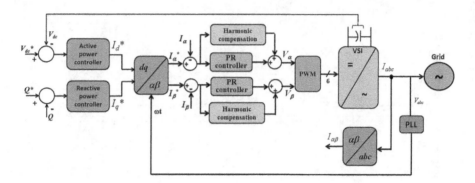

FIGURE 2.49 Grid-connected VSI control in stationary reference frame with PR controller.

where n_0, n_1, n_2, d_1, d_2 are numerator and denominator coefficients. Therefore,

$$y(k) = n_0 u(k) + n_1 u(k-1) + n_2 u(k-2) - d_1 y(k-1) - d_2 y(k-2) \qquad (2.55)$$

where $u(k)$ and $y(k)$ are the sampled input and the output signals of the discrete PR controller, respectively.

2.4.4 MODELING OF ACTIVE–REACTIVE POWER CONTROLLERS WITH DROOP CONTROL

2.4.4.1 Active Power Controller

As seen in Figure 2.24, the power balance controller receives DC link voltage as input and compares with its reference quantity for instantaneous power balance control. The DC capacitor voltage depends on the energy balance between the power received by the VSI and that delivered by it; only when these two are equal, the DC link capacitor voltage will remain constant as depicted in Figure 2.28. It thus ensures that the entire active power from the preceding stage is delivered to the grid.

The MPPT algorithms of RE generators will be used to provide the power reference to the power balance controller for the sake of maximum power tracking; the actual power delivered can be computed from the inverter output quantities in synchronous reference frame from equations 2.35 and 2.36. The active and reactive power references in microgrids will be suggested by a central control station intended to maintain the generation-demand balance within the local area.

Consider a grid-tied microgrid that injects power P_1 through a VSI at the instant t_1. Suppose the main grid fails at this instant and the VSI has to migrate to islanded mode. Let the main grid be at its rated frequency of ω_{rated} at t_1 and the frequency starts falling as shown in Figure 2.50, which is the active power-frequency droop characteristics of the VSI. The droop characteristics show how the VSI ramps its power output up to meet the local demand when the main grid goes off. Let the minimum allowable frequency is ω_{min} at which the VSI produces P_2 and it occurs at the instant of t_2. It means that the process of islanding is complete at t_2, beyond which no power is drawn from the grid and the entire demand on the microgrid is supplied by VSI.

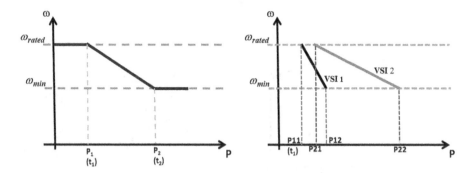

FIGURE 2.50   *P-f* droop characteristics of VSI.

Also, the VSI will deliver the power required by the local load at the rated frequency of ω_{rated} from the instant t_2 onwards. The droop characteristics thus help to decide the limit of local demand that can be met by the islanded microgrid. It also tells that (P_1-P_2) had been the power imported from the grid just before the islanding.

If the droop rate is constant and uniform, then the inverter will ramp up at uniform rate towards its rated power, P_{rated}; then the droop coefficient, m, is defined as,

$$m = \frac{\omega_{rated} - \omega_{min}}{P_{rated}} \tag{2.56}$$

Equation 2.56 assumes that P_2 equals P_{rated} of the inverter.

When multiple inverters are present on the microgrid at the instant of islanding, the smallest VSI (in terms of rated power) is preferred to move into grid-forming mode first. Once islanding is detected and grid is isolated, then the first VSI forms the grid and other VSIs get tied to it.

Therefore, whenever there is a deviation in frequency on the microgrid, then the power support by any VSI can be computed from equation 2.56 and it will serve as the reference in the respective active power control loop. This loop uses either a conventional PI controller or an advanced PR controller and forces the actual power to follow the reference power or force the DC link voltage to follow its reference as elaborated in section 2.4.3 and further in the later sections.

2.4.4.2   Reactive Power Controller

The reactive power controller seen in Figure 2.24 receives the reference quantities according to one or more of the following requirements as described earlier and presented again here: (i) The reactive power support to be provided by the converter, (ii) the power factor at which the current is to be delivered, and (iii) voltage control required on the AC grid. However, the reactive power reference should be set at zero if the inverter current is required to be delivered at unity power factor; this is often advocated in grid-feeding mode.

The controller can receive the reactive power reference (*Q*-reference) in any operating mode from the microgrid central controller, depending on the reactive power

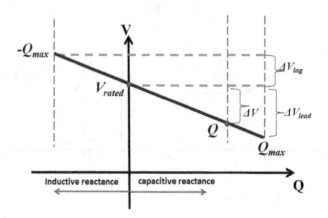

FIGURE 2.51   Q-V droop characteristics of VSI.

support to be provided by the VSI. The *Q-delivered* is the other input to the reactive power controller, which is calculated from the inverter output quantities in synchronous reference frame from equation 2.36. Often, a *Q-V* droop characteristic similar to that in active power control is adopted to accomplish AC voltage regulation on microgrid through reactive power control. It is designed to deliver a reference value to the inner current controller, be it inductive or capacitive reactive current. It can be noted from Figure 2.51 that a reactive power injection in an appropriate direction as ΔV_{lag} or ΔV_{lead} can cause a change in voltage, ΔV, in a required direction. The limit on the reactive power injection, Q_{max}, of Figure 2.51 is determined by the kVA rating of the VSI, which decides the maximum voltage sag or swell that can be corrected by the VSI. The Q-V droop equation is,

$$V = V_{rated} - m_Q Q_{inj} \tag{2.57}$$

where m_Q is the reactive power droop slope, Q_{inj} is the reactive power needed for voltage restoration, V is the required AC bus voltage and V_{rated} is the rated AC bus voltage. Therefore, whenever a voltage deviation is observed on the AC bus of the microgrid, an appropriate value of Q_{inj} will be calculated by the reactive power controller through Equation 2.57 and will be applied as Q-reference to this controller.

2.4.5   EMERGING NON-LINEAR CONTROLLERS

Non-linear controllers are robust which exhibit good dynamic response in reference tracking. Non-linear controllers developed in the past include model predictive controller, H-infinity controller, sliding mode controller, neural network-based controller, and fuzzy logic controllers. Model predictive controllers (MPC) have an edge over the others thanks to their multi-objective control through a single cost function optimization to achieve multiple targets. Modeling of MPC for grid-tied VSI is presented in detail in the following sections along with brief descriptions of control features and functionalities of other non-linear controllers.

2.4.5.1 MPC for Grid-Following Inverters

New age converter controllers are expected to accomplish additional capabilities such as seamless bi-directional power transfer, multiple control targets, adaptability to system non-linearities, faster dynamic response, etc., besides the prime objective of reference tracking. The MPC excels in simultaneous achievement of multiple control objectives of a heterogeneous nature.

A discrete time model of the system is used in MPC, to predict possible outcomes over a predefined time horizon. Any non-linearity in the system can be thus incorporated while formulating the prediction model. The optimal control is achieved by minimizing a cost function, g, that represents the desired target of the control,

$$g = f_m + \lambda_n f_n \tag{2.58}$$

where f_m is the hard constraint or primary target, f_n is the soft constraint or secondary target and λ_n is a weight parameter. In grid-following inverters, the reference current will be the primary target while the secondary target may be one or more of such features as harmonic profile improvement, switching frequency reduction, reactive power control, etc. Conventionally, both of these targets are attained in MPC through optimization of a single cost function. A firm minimization of f_m gives accurate reference tracking, while f_n will be graded relative to f_m by the weight parameter. The values of λ_n vary between 0 and 1 depending upon the priority of the auxiliary control implemented via the corresponding f_n.

The mathematical model of a three-phase grid-feeding inverter with MPC current control is presented here. The possible output currents are predicted for every possible inverter switching state. These predicted currents are passed on to the optimization process with the defined cost function. Optimization of the cost function will suggest the switching state of the inverter that will yield the smallest error between the reference and the actual value of current. Then the selected state will be applied to the inverter in the subsequent iteration.

2.4.5.1.1 Modeling of MPC Based Grid-Tied Inverters

The schematic diagram of MPC based grid-tied inverter system is shown in Figure 2.52, with V_{DC} as the input DC voltage, $i*(k)$ as the reference current of kth sample intended to be delivered by the inverter, and $i(k)$ as the corresponding actual current.

The load model of the grid connected inverter can be derived from the fundamental voltage equations obtained by applying Kirchhoff's voltage law at the inverter output,

$$v_{xN} = L\frac{di_x}{dt} + Ri_x + v_{Gx} + v_{nN} \tag{2.59}$$

where v_{xN} ($x = a$, b, c) are the inverter voltages, v_{Gx} are the grid phase to neutral voltages, v_{nN} is the voltage between load neutral and DC bus ground, i_x are the phase currents of inverter, R is the filter resistance, and L is the filter inductance.

Equation 2.59 can be represented in space vector form as,

$$v = L\frac{di}{dt} + Ri + \hat{v}_G \tag{2.60}$$

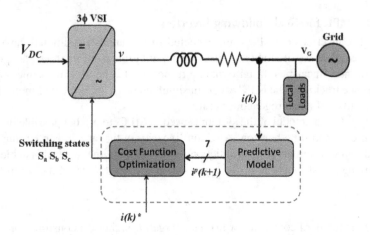

FIGURE 2.52 Three phase grid-tied inverter with MPC.

where v is the space vector of inverter output voltage, $\hat{v}_G$ is the peak value of grid voltage and i is the load current vector. Alternatively, the inverter voltage vector can be expressed as,

$$v = V_{DC}(s_a + as_b + a^2 s_c) \tag{2.61}$$

where $a = e^{j2\pi/3}$, and, s_a, s_b and s_c represent the status of the top switches of the inverter legs. Thus, a three-phase inverter with six switches will have eight possible voltage vectors represented as v_0 to v_7, from the inverter states 000–111.

Converting equation 2.60 into its discrete time model by applying Euler's approximation with a sampling time of T_s gives,

$$\frac{di}{dt} \approx \frac{i(k+1) - i(k)}{T_s} \tag{2.62}$$

$$v(k) = L\left(\frac{i(k+1) - i(k)}{T_s}\right) + Ri(k) + \hat{v}_G(k) \tag{2.63}$$

$$i(k+1) = \left(1 - \frac{RT_s}{L}\right)i(k) + \frac{T_s}{L}\left(v(k) - \hat{v}_G(k)\right) \tag{2.64}$$

Now, the eight possible values of the future current, $i^p(k+1)$ will be predicted as,

$$i^p(k+1) = \left(1 - \frac{RT_s}{L}\right)i(k) + \frac{T_s}{L}\left(v(k) - \hat{v}_G(k)\right) \tag{2.65}$$

Equation 2.65 requires the grid voltage value at kth sample. However, this can be estimated by back extrapolation. The estimated peak value of grid voltage can be expressed as,

$$\hat{v}_G(k-1) = v(k-1) - \frac{L}{T_s}i(k) - \left(R - \frac{L}{T_s}\right)i^p(k-1)$$

(2.66)

The back extrapolation estimates the grid voltage for the preceding sample with the assumption that the grid voltage does not vary within the sampling interval. Therefore, the estimated $\hat{v}_G(k-1)$ is applied in place of $\hat{v}_G(k)$ in Equation 2.65 to complete the current prediction for $(k+1)$th sample which is then sent for optimization.

2.4.5.1.2 Formulation of Cost Function

The cost function can be developed in any reference frame and it does not alter the tracking capability of MPC. The cost function, g, of the grid-tied inverter developed here to track a current reference in stationary reference frame without any secondary targets is expressed as,

$$g = \left|i_\alpha^* - i_\alpha^p\right| + \left|i_\beta^* - i_\beta^p\right|$$

(2.67)

where i_α^*, i_β^*, i_α^p and i_β^p are the stationary reference frame coordinates of the reference current, i^*, and the predicted current, i^p, respectively.

Eventually, the α and β current components are predicted at every sampling instance using equations 2.65 and 2.66 and MPC is executed to switch the inverter.

2.4.5.1.3 Performance Evaluation of MPC-Based Grid-Tied Inverter

A three phase grid-tied inverter with specifications given in Table 2.2 is simulated and controlled using MPC and its performance is presented in this section. The cost function presented in Equation 2.67 is used to track a rms current of 15.19 A and inject 14.8 kW of power to the grid. The formulated MPC controller is sampled at 25 kHz.

The current delivered by the inverter in Figure 2.53b shows a close compliance with the reference current of Figure 2.53a confirming the tracking ability of the MPC inverter. The current vector trajectory of Figure 2.53e is the plot of the reference and the actual current of the grid-tied inverter system. Inverter current (i) and reference current (i^*) exhibit a high degree of agreement with each other except minor deviations. The switching frequency of the inverter will be varying while tracking the reference as it is a current regulated PWM.

TABLE 2.2
Inverter Specifications

Parameters	Value
DC source voltage	700 V
Inverter VA rating	15 kVA
Grid specifications	3Φ,400 V(l-l), 50 Hz
Filter inductance	0.001 Ω, 10 mH

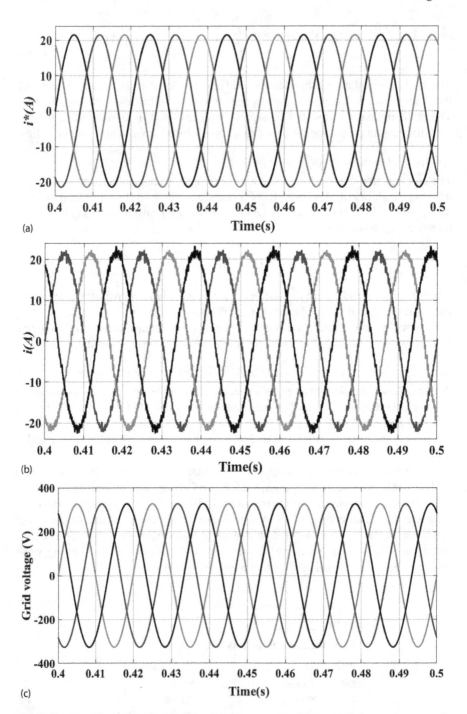

FIGURE 2.53 The MPC based grid-tied inverter performance: (a) Reference current, i^*, (b) current delivered by inverter i, (c) grid voltage.   (*Continued*)

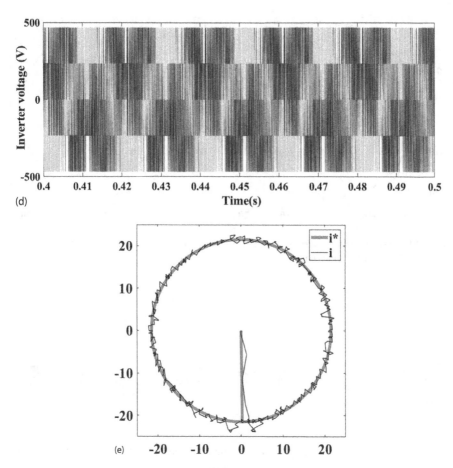

(d)

(e)

FIGURE 2.53 (Continued) The MPC based grid-tied inverter performance: (d) inverter leg voltage, and (e) current vector trajectory of I^* and I in stationary reference frame.

As the switching frequency and the model prediction depend on the sampling frequency, f_s, used in MPC, the tracking performance depends greatly on the sampling frequency. Therefore, the system is tested for sampling frequencies of 50 and 100 kHz, with the same reference currents and the corresponding current vector trajectories are presented in Figure 2.54.

Figure 2.55a–c depict the line current error, i_{aerr}, plotted against samples. The current error per sample is found to get reduced with increase in f_s. Figure 2.55d shows the frequency spectrum of the delivered current, which is found to have harmonics well within the stipulated limits. The harmonic spectrum with any f_s is widely spread up to half of f_s. Such spread spectrum can be shaped with the addition of a secondary target.

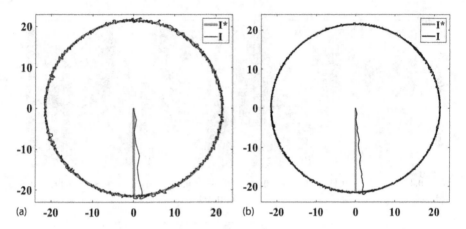

FIGURE 2.54 (a and b) Current vector trajectory of $i*$ and i in stationary reference frame with $f_s = 50$ and 100 kHz respectively.

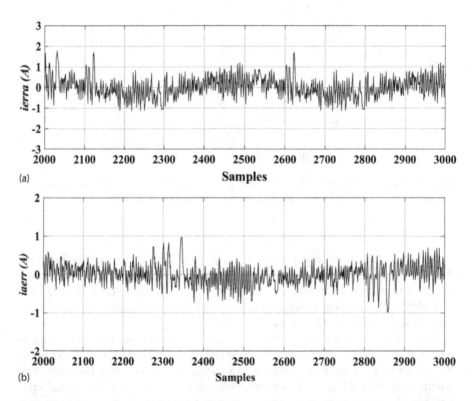

FIGURE 2.55 (a–b) Instantaneous current error, i_{aerr}; with f_s of 25, 50 and 100 kHz respectively.   (*Continued*)

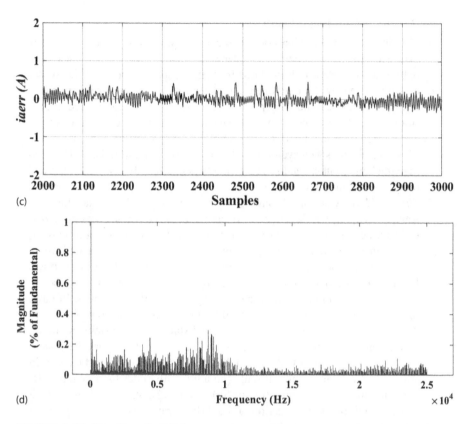

(c)

(d)

FIGURE 2.55 (Continued) (c) Instantaneous current error, i_{aerr}; with f_s of 25, 50 and 100 kHz respectively, (d) Frequency spectrum of current for 50 kHz.

Tracking accuracy and THDi are the performance indices considered. Tracking accuracy is defined in terms of magnitude, A(mag), and phase angle, A(angle), and expressed in % as,

$$A(\text{mag}) = \left(1 - \left(\frac{I^*_{rms} - I_{rms}}{I^*_{rms}}\right)\right) \times 100 \qquad (2.68)$$

$$A(\text{angle}) = \left(1 - \left(\frac{\theta^* - \theta}{360°}\right)\right) \times 100 \qquad (2.69)$$

where I^*_{rms}, I_{rms}, θ^* and θ are the rms values and phase angles of reference current and inverter delivered current, respectively. The values of tracking accuracy obtained in the test case for both magnitude and phase angle are well above 99.5%. These performance merits clearly show that MPC is a competent contender for the grid connected converter control. It is identified as a prominent alternative to voltage based classical PWM techniques as well as other implicit modulator techniques.

2.4.5.2   Other Non-linear Controls

1. *H-Infinity Controller*: H∞ control is adopted when robust performance is expected in spite of system parameter variations and large disturbances. An optimization process is formulated from the problem which will be subsequently solved by the controller. The design requirements like disturbance rejection, robustness, tracking performance, etc., are to be formulated as constraints in different control loop transfer functions. The weighting functions are selected so as to tune these loops until the desired performance is reached. This control works well even with unbalanced load, exhibits reduced THD and high tracking accuracy and is easy to implement. Slow dynamic response and requirement of multiple control loops are the disadvantages of H∞ controller.
2. *Sliding Mode Control*: A sliding mode controller (SMC) has inherent robustness against wide range of system parameter deviations, external disturbances even with their strong uncertainties. If any plant response deviates from its normal operating points, a discontinuous control will direct the system's state trajectories to persist on some desired sliding surface. Because of the discontinuous regulator, a strong control action occurs, which provides an excellent dynamic response for the controller. In basic SMC, a discrete control law is defined for the system under control with the desired performance defined as the desired states of the system. A sliding surface and the switching conditions will be such defined that the system states are made to follow the desired states. A good sliding surface in grid current control can ensure current regulation with better harmonic profile. A commonly reported limitation of SMC is the chattering problem, which is rectified through optimization of the SMC parameters and by addition of integral terms onto the sliding surface to eliminate tracking errors.

2.4.5.3   Artificial Intelligence-Based Current Controllers

Artificial intelligence-based controllers have been developed for various applications that employ VSI. Neural network and fuzzy logic controllers are two such tools adapted in these current controllers.

- *Neural Network Controller*: Neural network (NN) controller falls under the category of non-linear controllers. Learning ability of the controller makes it an obvious choice for microgrids having uncertain system models with wide range of parameter variation induced by RE source intermittency. Dynamic programming can be used to train the NN and the training is offline. It is possible to implement the real-time control action with the modern high speed-large memory microcontrollers without significant delay and with low computing power requirements. The NN controllers show fast operation and good dynamic response especially in MPPT control under fast changing ambient conditions. Though online optimization is a constraint in NN controllers, it can be overcome by adopting parallel processing architecture.
- *Fuzzy logic current controller*: Fuzzy logic controller is a replacement for the traditional PI controllers to work in non-linear dynamic systems. Fuzzy logic methodology handles non-linear dynamics effectively, as it is

a model-free approach using the designer's knowledge base to fine tune the controller actions by simple *If-Then* rules. A fuzzy logic controller in grid-tied VSI receives the inverter current error and its derivative as inputs and it delivers the desired reference voltage, with the help of the knowledge base, to the PWM generator. Further, fuzzy controllers are also effective in handling the microgrid model dynamics, especially in transient conditions..

2.4.6 CONTROL UNDER AUTONOMOUS MODE

Microgrids are expected to operate in grid-forming mode upon islanding and subsequent isolation from the main grid. The inverter in this mode is controlled to *form* the grid by establishing the rated voltage and frequency in the local grid. The control objective is to regulate the amplitude and frequency of the voltage at the inverter output and provide the active and reactive power demand of the loads on the microgrid. The grid-forming inverters are generally powered from storage battery. However, in recent times, inverters of various RE sources are also entrusted with the responsibility of grid formation, yet with adequate transient power support provided by fast responding storage systems like super capacitors and battery. Accomplishing grid formation with PV inverters is relatively easier than that with the inverters of wind electric systems. Grid-forming converters are to be enabled with black start capability, as this mode of operation starts instantaneously with islanding.

2.4.6.1 Voltage and Frequency Regulation in Grid-Forming Mode

The grid-forming power converters are controlled in any frame of reference like SRF-dq or stationary reference frame-$\alpha\beta$, depending on the computing power available. Various methods are available to generate the reference signal for the grid-forming controller, like look-up-table, local oscillators, etc. The control scheme for a grid-forming converter is presented in Figure 2.56, wherein the reference voltage, V^*_{rated}, and the reference frequency, ω^*_{rated}, are pre-fed into the controller and these represent the voltage and frequency to be established at PCC. The outer loop regulates the voltage at PCC by maintaining the charge balance of the filter capacitor, C_o. The holdup time in

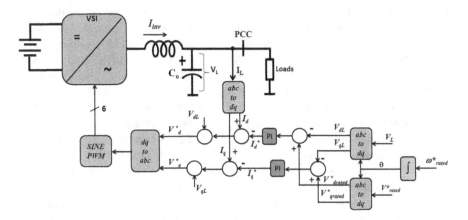

FIGURE 2.56 Grid-forming control in SRF.

equation 2.9, when applied in grid-forming applications to design C_o, has to consider the source power variation as well as the load dynamics. The variation in the voltage at PCC, V_L, is an indication of mismatch between the current delivered by the inverter and that demanded by the load. The inner loop regulates the current so as to maintain the voltage across C_o as V^*_{rated}. The dynamic control equation for voltage regulation is,

$$\frac{dV_L}{dt} = -\frac{1}{C_o}(i_{inv} - i_L) \tag{2.70}$$

where I_{inv} and I_L are the inverter current and the load current, respectively. The dynamics of the voltage control loop in dq reference frame can be expressed as,

$$C_o \frac{dV_{Ld}}{dt} = C_o \omega V_{Lq} + i_{invd} - i_{Ld}$$

$$C_o \frac{dV_{Lq}}{dt} = -C_o \omega V_{Ld} + i_{invq} - i_{Lq} \tag{2.71}$$

where V_{Ld} and V_{Lq} are the d and q axes components of the voltage at PCC, ω is the frequency at PCC, i_{invd} and i_{invq} are the d and q axes components of the inverter current, i_{Ld} and i_{Lq} are the d and q axes components of the load current, respectively.

Grid-forming mode control will make the inverter interact with the loads in the microgrid in the absence of the main grid as well as any local synchronous generation. So, black start capability is essential in this mode; in other words, there must be a reliable source to initiate the formation of the microgrid from a complete shutdown. Additionally, the grid-forming asset is expected to dispatch a required amount of power to the microgrid loads. The voltage generated by the grid-forming inverter will serve as the reference for the rest of the grid-feeding inverters of the microgrid. Deferrable loads, diesel generators and high-power density storages like super capacitors in conjunction with the battery can improve the transient power support within the microgrid which eventually improves the grid reliability in autonomous mode.

2.4.6.2 Black Start Capability

Black start can be defined as the capability of the microgrid to boot up without any external power support. Auxiliary battery powered systems can provide black start in a microgrid. The black start requirement of a microgrid can be classified based on the operating condition; namely, whether it is a *cold start* (a first time start up) or a *re-connection* (after a fault or an intentional shut down). The "grid former" is expected to ensure short start-up time so that fast reconnection of loads is possible. The asset which takes the responsibility of grid formation should also be rated to handle large in-rush currents during the re-connection of typical loads.

Sometimes, diesel generators are assigned the responsibility of black start so that the inverters in the microgrid can synchronize to it. However, the black start time required by a diesel generator is high (typically 10–20 s), so it hardly fits the definition of uninterrupted power supply. Moreover, diesel generators do not find a place in 100% RE microgrids. So, a battery-powered inverter has to play the role of grid former or master generator in such systems; it provides faster post-fault reconnection unlike the diesel generator. Yet, it is ideal in autonomous RE microgrids to assign one of the RE generators to be the grid former and either a battery or super capacitor can be assigned as the standby. The RE fed master generator will have a hybrid inverter capable of moving into grid-forming mode in case of islanding. The mode change is possible as long as the RE source has energy to maintain a stiff DC-link at the inverter input; else, the standby battery will replace the RE generator.

There is, however, an interesting question as to which RE generator should be given the responsibility of grid formation. There are two contradicting approaches: The first being to optimize the power rating of the master generator in order to reconnect as many loads as fast as possible, and the second to focus on the maximum utilization of the RE source through MPPT. The voltage regulation required in the grid-forming mode often will not support MPPT, and as a consequence choosing a large inverter for grid tying will not guarantee higher utilization of RE. If a larger number of inverters are used to form the grid (using an appropriate parallel operation algorithm during the mode change from grid-tied to grid-forming), then rapid reconnection of a large number of loads can be possible; this approach can also reduce the power loss due to off-MPPT operation.

Further, the stochastic nature of RE power will create an additional issue of intermittency in the grid-forming operation. This can however be addressed by addition of dynamic energy storage systems like super capacitors which can impart transient stability against RE variability.

2.4.6.3 Grid Forming with Wind Driven Generators

Wind electric generators will have a major share of distributed generation in microgrids. Variable speed wind electric generators like permanent magnet synchronous generators (PMSG) and doubly fed induction generators (DFIG) have proven capability to extract more from wind energy than fixed speed squirrel cage induction generators (SCIG). Of late, the wind electric generators are in demand to operate in autonomous mode on autonomous microgrid. The following sections address the microgrid interfacing and control of PMSG and DFIG in autonomous mode or grid-forming mode of operation.

2.4.6.3.1 Grid-Forming Control of PMSG

The PMSG is preferred to SCIG and DFIG in stand-alone wind energy systems because of the self-excitation characteristic. Figure 2.57 presents the grid forming

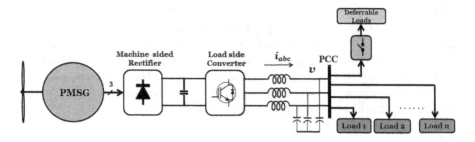

FIGURE 2.57 Wind driven PMSG in autonomous mode.

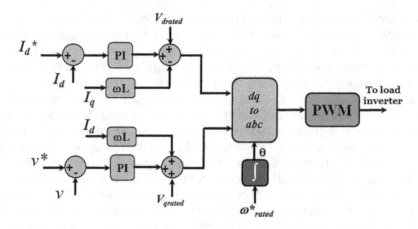

FIGURE 2.58 Grid-forming control of wind driven PMSG system.

scheme of wind driven PMSG in autonomous mode, where the load side inverter is controlled to establish the grid at PCC. However, the power variance due to intermittency in wind speed and its implications on the AC voltage regulation are very severe which demand complex control of the inverter. The load side inverter control of the PMSG system for grid formation is presented in Figure 2.58.

This PMSG system has only one controllable converter, which is the load side inverter. So, the grid formation control targets the output voltage and frequency of the load side inverter using SRF controllers explained in Section 2.4.3.3. The control scheme of Figure 2.58 follows the SRF control voltage equations from 2.30 to 2.34, and targets delivery of the active and reactive powers of equations 2.37 and 2.38 to the loads. The voltage regulation is obtained through the outer loop of the reactive power controller, wherein the rated load voltage is provided as a reference and compared with the measured voltage at PCC. The rated frequency is used to obtain the sine reference for the PWM block of the load side inverter of Figure 2.58. This control does not account for unbalance in the load, power intermittency due to wind speed variations and MPPT operation of the wind turbine (WT). Therefore, alternative schemes are suggested here to address the aforesaid issues by adding a battery in the DC link and an additional DC-DC converter after the rectifier.

2.4.6.3.2   MPPT Operation of WT

The main advantage of a variable speed wind electric generator is that the average aerodynamic efficiency of WT is relatively high, as the latter is made to operate at or near its maximum efficiency point by allowing the shaft speed to vary over a wide range in proportion to wind speed variation. This scenario is explained below.

The power input to the electric generator, or the mechanical power output of the wind turbine, can be expressed in Watts as,

$$P_m = \frac{1}{2}\rho A v_w^{\,3} C_P(\lambda, \beta) \qquad\qquad (2.72)$$

where C_P is the power coefficient which is a function of the tip speed ratio, λ, and blade pitch angle, β; ρ is air density in kg/m³, A is the blade swept area in m² and v_w is the wind velocity in m/s. λ is the ratio of blade tip speed (which is the product of angular speed of the turbine, ω_T, and blade length, R) to v_w. When the wind electric generator operates in varying wind speed, λ will vary if the variations in ω_r and v_w are disproportionate; then the C_P variation will be hyperbolic and its maximum value, $C_{P\max}$, will correspond to optimum tip speed ratio, λ_{opt}, for a given β.

Therefore, the mechanical power extracted by the WT when operates at $C_{P\max}$ is,

$$P_{m,\max} = \frac{1}{2}\rho A v_w{}^3 C_{P\max}(\lambda,\beta) \tag{2.73}$$

and the corresponding shaft speed is,

$$\omega_{Topt} = \frac{v_w\,\lambda_{opt}}{R} \tag{2.74}$$

In order to ensure maximum power point tracking in a wind electric generator, the shaft speed is forced to vary continuously such that λ is kept constant at λ_{opt} against any variation in v_w; this is called the TSR-MPPT method. An optimum torque MPPT (OT-MPPT) algorithm can be further developed continuing on the same concept as follows. The mechanical torque delivered by the WT can be obtained from equation 2.72 as,

$$T_m = \frac{P_m}{\omega_T} = \frac{\frac{1}{2}\rho A v_w{}^3 C_P}{\omega_T} \tag{2.75}$$

If the turbine shaft speed is maintained at ω_{Topt}, then the optimum value of torque to be obtained from WT can be identified and subsequently used as the reference in the control loop. By use of equations 2.73–2.75 can be rewritten as,

$$T_{mopt} = \frac{1}{2}\frac{\rho\pi R^5 C_{P\max}\omega_{Topt}^2}{\lambda_{opt}^3} = K_{opt}\omega_{Topt}^2 \tag{2.76}$$

Computing T_{mopt} from equation 2.76 with the sensed shaft speed, the reference for the MPPT control loop can be obtained.

MPPT is not possible in stand-alone mode of operation, unless the demand matches the maximum power available; therefore, energy storage is integrated into the stand-alone WT driven PMSG system for harnessing maximum power. And the stored energy can be used to compensate the power deficit during intermittent dips in wind speed. Such a scheme of WT driven PMSG is presented in Figure 2.59.

The MPPT converter in Figure 2.59 is a DC-DC converter which is controlled to extract maximum power from the WT-PMSG. The MPPT control loop senses the shaft speed and ensures that the sum of the power delivered by the load side converter and the power delivered to battery is as demanded by the shaft speed such that λ_{opt} is maintained in the WT. In addition to the storage, deferrable loads can

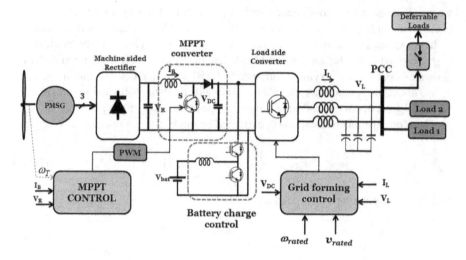

FIGURE 2.59 WT driven PMSG in autonomous mode with MPPT control and battery backup.

be identified on the AC bus in order to ensure MPPT even when the battery is full. Deferrable loads are loads which can wait until power is available in excess of the regular load demand, which means these need be dispatched on a fixed schedule. The presence of energy storage and deferrable loads will increase RE utilization and make the microgrid operation more flexible. The control schemes for TSR-MPPT and OT-MPPT are presented in Figure 2.60.

A separate DC-DC converter establishes battery charge control and supports MPPT operation of WT by constituting itself as a controllable load. The combined load on WT comprises the regular load on the AC bus, the battery and the deferrable load. At any instant one or more of these three loads are regulated to maintain the shaft speed of equation 2.74. Despite the varying demand on the AC bus, the battery

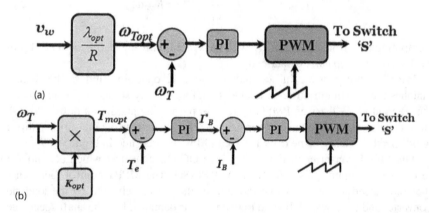

FIGURE 2.60 MPPT control for WT-PMSG: (a) TSR-MPPT and (b) OT-MPPT.

charge controller is operated to adjust the charging current to fill the gap between the total generation and demand. Such matching of generation to demand is essential to ensure the operation of WT at its λ_{opt} and C_P at its maximum value. Upon the battery reaching its maximum capacity – with the requirement to sustain the match between total generation and total demand – the deferrable loads are operated.

The grid-forming control of grid side converter of PMSG in autonomous mode is presented in Figure 2.61. SRF controller is used to accomplish voltage and frequency regulation. The voltage and frequency references represent the respective rated values. In Figure 2.61, imparting the value of reference frequency, ω^*_{rated}, will provide the value of θ needed for abc to dq transformation and will ensure the inverter output voltage will have this frequency. The DC link voltage regulation will provide the active power reference of the inverter as discussed in Section 2.4.1.3.1. The load voltage at PCC is measured and its d-axis component, V_d is regulated with an outer PI regulator to obtain the current reference I_d^*. V_{drated} in Figure 2.61 represents the d-axis component of the required AC bus voltage which is added along with the PI compensated filter drop and the cross coupling factor to obtain the d-axis reference voltage V_{di}^* for the inverter as per equations 2.30–2.34. Similarly, in the reactive power loop, the amount of i_q^* to be delivered along with the cross-coupling factor amounts to be the q-axis reference voltage for the inverter V_{qi}^*. These reference voltages will accomplish the required AC bus voltage through PWM switching of the load side inverter. In addition, the maximum value for the DC link voltage, V_{DCmax}, can be declared and any increase beyond this value is an indication of excess power available in the system, and then the deferrable load can be operated for power balance.

2.4.6.3.3 Grid-Forming Control of DFIG

The DFIG is the cost-effective choice of generator in large variable speed WT systems as the power electronics carries only the slip fraction of the rated power.

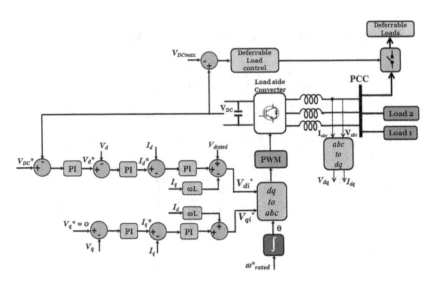

FIGURE 2.61 Grid-forming control of load side converter of DFIG.

But its range of operating speed is also limited, by the slip, unlike in PMSG. Its inherent double feeding feature provides the opportunity to integrate energy storage into its rotor circuit and it is controlled to accomplish different operating conditions of the generator. However, the stand-alone operation of DFIG demands yet more sophisticated control as the stator voltage and frequency regulation has to be obtained through a rotor side converter. Several control methods are available for grid-forming operation of DFIG including direct voltage control, field-oriented control, direct power control, etc.

- *Concept of voltage and frequency regulation in DFIG*: The steady state equivalent of a slip ring induction machine operating in stand-alone mode and supplying a load of $R + jX$ is presented in Figure 2.62.
  Here V_r is the rotor voltage, E represents the air-gap voltage, a is the turns across ratio the stator and rotor, I_s is the stator current, I_r represent the rotor current referred from stator, V_s is the stator terminal voltage, R_r is the rotor resistance referred to stator, X_r is the rotor reactance referred to stator, R_s and X_s are stator resistance and reactance respectively, R_m and X_m are shunt resistance and magnetizing reactance respectively, s is the slip. The KVL equation of the rotor side circuit results,

$$a\frac{V_r}{s} = E + I_r\left(\frac{R_r}{s} + jX_r\right)$$ (2.77)

The air gap voltage E in terms of stator voltage and the stator drop be expressed as,

$$E = V_s + \frac{V_s}{(R+jX)}(R_S + jX_S)$$ (2.78)

The rotor current due to the applied rotor voltage is,

$$I_r = \frac{V_S}{R+jX} + V_S\left(1 + \frac{R_S + jX_S}{R+jX}\right)\left(\frac{1}{R_m} + \frac{1}{jX_m}\right).$$ (2.79)

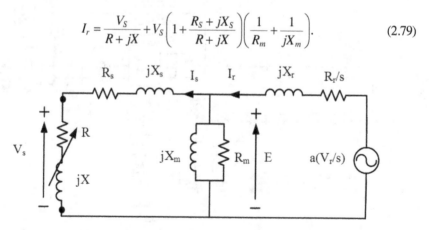

FIGURE 2.62   Equivalent circuit of DFIG in stand-alone mode.

On substituting equations 2.78 and 2.79 in equation 2.77 and simplifying to result the stator voltage is,

$$V_S = a\frac{V_r}{s}\frac{1}{\left(1+\dfrac{1+(R_S+jX_S)}{(R+jX)}+\left(1+\dfrac{R_S+jX_S}{R+jX}\right)\left(\dfrac{1}{R_m}+\dfrac{1}{jX_m}\right)\left(\dfrac{R_r}{s}+jX_r\right)\right)} \qquad (2.80)$$

Equation 2.80 conveys that in stand-alone mode of DFIG, the stator voltage magnitude varies with variations in slip, the rotor voltage and the load on the stator. Thus, whenever the load or the slip varies, the control should counteract by varying the rotor voltage such that the stator voltage is maintained at its rated value.

Similarly, the stator frequency of DFIG when working in stand-alone mode varies due to the generation-load mismatch on the system. When the power available from the generator is more than the load connected on the stator, then the shaft speed increases and while the power generated is less than the load the shaft speed decreases. During this speed transition, if the shaft speed is greater than the synchronous speed, then the mode of operation is called super-synchronous while a lower shaft speed than synchronous speed results a sub-synchronous operation. Such speed variations cause the magnetic field of the machine to rotate at speeds other than the synchronous speed, thus deviating the stator frequency from its rated value. In stand-alone DFIG, if a constant stator frequency is to be established and sustained then the air gap magnetic field needs to be rotating at synchronous speed.

The relationship between the stator and rotor frequency in induction machines can be written as,

$$\omega_r = s\omega_s = \omega_s - \omega_m\frac{P}{2} \qquad (2.81)$$

where ω_r is the rotor frequency in rad/s, ω_s is the stator frequency rad/s, ω_m is the mechanical angular speed, and P is the number of poles. In equation 2.81, if ω_m varies due to change in load or change in the wind speed, then the frequency injected at rotor circuit should be varied so as to maintain ω_s, at its rated value. So, the desired rotor frequency to be injected can be obtained from equation 2.81. Such a variable frequency variable voltage can be obtained through an inverter in the rotor circuit, which can be powered by a battery source or by a DC grid in case of a hybrid microgrid of Figure 2.6 dealt in Section 2.2.1.1.3.

With reference to Figure 2.16, DFIG can be made to work as generator in both super-synchronous and sub-synchronous modes of operation. In sub-synchronous generation, the active power flows into the rotor circuit, whereas the active power is delivered by the rotor circuit in super-synchronous generation. Therefore, the converter provided in the rotor circuit has to support bidirectional power flow, simultaneously injecting a voltage with appropriate magnitude and frequency.

The rotor in any induction machine has induced voltage at slip times the stator frequency which creates its own magnetic field rotating at the rotor frequency. Since the wind turbine is rotating the DFIG rotor shaft, the speed of rotation of the rotor magnetic field is together decided by the injected rotor frequency and the mechanical rotation of the rotor shaft. This rotor magnetic flux linking with the stator windings and the rate of change of that flux linkage will decide the magnitude and frequency of the stator voltage.

The sub- and super-synchronous generation by DFIG will have its rotor shaft speed below and above the synchronous speed respectively. This will respectively result in positive and negative values of ω_r as per equation 2.81. The change in polarity of ω_r indicates the change of phase sequence of the three phase voltages. This phase sequence reversal is necessary to facilitate the reversal of power in the rotor circuit while the operation moves across sub- and super-synchronous modes.

A variety of stand-alone DFIG controllers are reported in research literatures and are broadly classified as (i) sensorless controllers and (ii) controllers with rotor position encoder. The stator voltage magnitude at a given speed and load will be proportional to the rotor current and its frequency. Once the rotor current frequency is decided based on Equation 2.81, then the magnitude of rotor current is varied until the reference amplitude of the stator voltage is generated.

- *Stator voltage and frequency control with speed encoder*: The block diagram of simple stator voltage and frequency control to form a microgrid with DFIG is presented in Figure 2.63. This scheme includes a speed encoder and utilizes the position information in the control loop to establish the frequency relationship of equation 2.81. The reference angle of the rotor current i_r is obtained by subtracting the angle corresponding to the

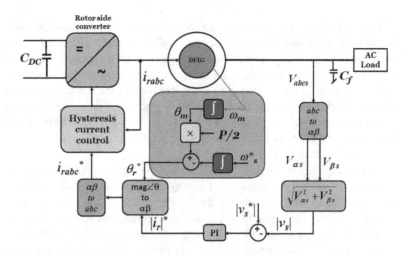

FIGURE 2.63 Grid-forming control of DFIG with speed sensor.

rotational speed, θ_m, from the angle corresponding to the reference stator frequency, θ_s, as,

$$\theta_r^* = \int \omega_s^* dt - \frac{P}{2}\theta_m \qquad (2.82)$$

where ω_s^* is the reference stator frequency. The reference space vector magnitude of stator voltage, $|v_s^*|$, is applied and the actual stator voltage, $|v_s|$, is calculated using the orthogonal components in the stationary $\alpha\beta$ reference frame. The load changes contribute to the error between these two voltages, and so the compensated error serves as the reference rotor current magnitude, $|i_r|^*$. This is fused with the angle information (θ_r^*) of equation 2.82 to complete the rotor reference current vector calculations. This current vector is further transformed as three phase current references, i_{rabc}^*, and is used in a hysteresis current controller to make the actual rotor current to follow this. Conversely, these rotor reference currents can also be used with SRF current controllers, instead of the hysteresis controller, as described in Section 2.4.3.2. Capacitor, C_f, at the stator terminals can serve as the filter as well as provide reactive power support for the stand-alone loads.

- *Stator voltage and frequency control without speed encoder:* The angle of the reference rotor current if obtained without a speed encoder represents a sensorless DFIG system. θ_r^* can be obtained by the principle of phased locked loops (PLL) adapted in grid-connected systems for synchronization. The PLL is used for the synchronization of the induced stator voltage vector with the arbitrarily assigned reference voltage vector. A small modification in the control loops along with a PLL can establish the stator voltage exactly in synchronism with the phase angle of the reference frequency. Such a sensorless grid-forming control of DFIG is presented in Figure 2.64, wherein the stator voltage is sensed and converted with SRF as dq quantities with the angle information obtained through integration of the reference stator frequency ω_s^*. The d- and q-axis voltages are used to obtain the magnitude of the stator voltage vector, $|v_s|$ and its angle θ_s as follows:

$$|v_s| = \sqrt{V_{ds}^2 + V_{qs}^2} \qquad (2.83)$$

$$\theta_s = \tan^{-1}\left(\frac{V_{qs}}{V_{ds}}\right) \qquad (2.84)$$

Two PI controllers are used to compensate the error in the voltage and the angle control loops. The reference for the voltage loop will be the magnitude of the rated stator voltage vector, $|v_s^*|$. The compensated voltage error will result the rotor current magnitude to be injected.

The angle reference, θ_s^*, is maintained as zero for the grid-forming control. This will ensure that the reference voltage vector aligns with the d-axis of the SRF. Such an alignment will result V_{qs} equal to zero and $V_{ds} = |v_s^*|$. Finally, the angle loop error after compensation will yield the reference rotor frequency, ω_r^*, and its integration will provide the angle of the rotor reference current, θ_r^*.

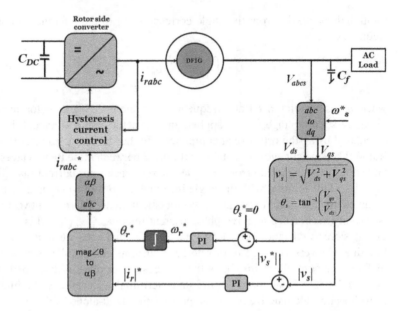

FIGURE 2.64   Sensorless grid-forming control of DFIG.

This angle, when fused with the rotor current magnitude information $|i_r|^*$ from the voltage loop, gives the reference rotor current vector. Further transformation of this current vector into three phase current reference, i_{rabc}^*, will complete the calculations. Subsequently, a hysteresis current controller is employed to make the actual rotor current to follow the reference current, i_{rabc}^*.

2.4.7   REACTIVE POWER SUPPORT IN MICROGRID

The autonomous microgrid has to maintain both active and reactive power balances within the network in the absence of the main grid. Active power management depends on the available generation and its control to match the load demand as discussed in Section 2.4.1.3.1. Reactive power compensation is necessitated due to the varying non-unity power factor loads and magnetization currents of transformers and induction generators as well as their influence on the microgrid voltage fluctuations. Non-linear loads are also a cause for poor power factors and large reactive power demand. Reactive power, or VAR, compensation is essential to improve the microgrid system performance in terms of voltage stability of its buses and improvement in transmission capacity within the network.

Several options are available for VAR compensation in microgrids. The converters in the microgrid can be controlled to deliver any desired combination of real and reactive powers; the droop P/V and Q/V droop concepts explained in Section 2.4.4. are predominantly utilized in autonomous mode of microgrids. However, it may not be possible to have unconstrained amount of VAR compensation while the converter is delivering active power close to its rated value. In other words, when the entire range of load VAR has to be compensated through the RE converters, these need to be oversized by several folds. It therefore demands a dedicated VAR compensation in microgrids as follows.

- *Static VAR compensator (SVC)*: The simplest of the solutions for VAR compensation is the *automatic power factor correction* with switched capacitor banks. The VAR can be better compensated by shunt-connected static VAR compensator (SVC) comprising thyristor controlled reactor (TCR) and thyristor switched capacitors (TSC). The SVC can provide precise VAR compensation with appropriate feedback control. A typical shunt compensation scheme with SVC is shown in Figure 2.65. Harmonics injection, resonance issues, grid voltage dependence and slow response make them inferior compared to other compensation techniques.
- *Static Compensator (STATCOM)*: One of the most effective VAR compensation technique is static compensator (STATCOM), a controlled power electronic device with fast dynamic response which can provide precise compensation. The STATCOM can be placed in selected locations on the microgrid and can be activated when necessary. It injects a variable magnitude sinusoidal current at PCC. It has a voltage source inverter tied to the grid through a shunt-connected transformer as shown in Figure 2.66a and its operating characteristics are shown in Figure 2.66b. The AC side current, i_{comp}, can be made lagging or leading with respect to its terminal voltage, V_{sta}, by PWM control of the inverter.

  The currents are injected in quadrature with the grid voltage, so STATCOM can realize either a capacitive or inductive reactance of adjustable value. The SRF power controllers described in Section 2.4.3.3 can be adopted here with only the reactive power loop. The reference Q can be obtained as per the load requirement, while the actual Q is calculated from the voltage and current at the inverter terminals.

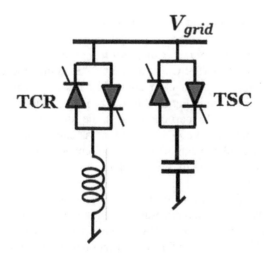

FIGURE 2.65 TCR and TSC.

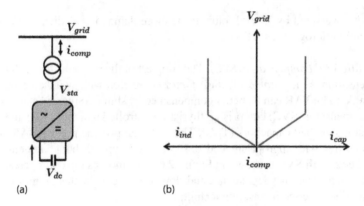

(a)   (b)

FIGURE 2.66 (a) STATCOM schematic and (b) operating characteristics.

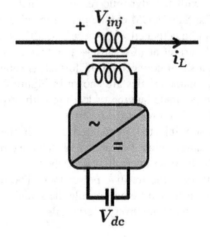

FIGURE 2.67 Static synchronous series compensator.

- *Static Synchronous Series Compensator (SSSC)*: Static synchronous series compensators (SSSC), is another VAR compensation technique. It has a voltage source inverter connected to the grid through a series transformer as shown in Figure 2.67a. It provides series compensation by injecting a voltage, V_{inj}, perpendicular to the line current, i_L, and leading it; this will maintain the injected voltage as capacitive. The magnitude of the injected voltage can be adjusted independent of the magnitude of the line current, thus realizing a variable capacitive reactance for exact compensation.
- Unified power flow controllers (UPFC), D-STATCOM, etc., are also technical options for VAR compensation in microgrid environment – choice of the method and system of VAR compensation depend greatly on the specific load pattern, amount of compensation required, other ancillary services required and most importantly, the cost.

2.4.8 GENERATION CONTROL

Power balancing is a major task in the operation and control of microgrid, as is already implied in the previous sections. Maintaining power balance and stability of a local network in the absence of the main grid is a challenge, unless the RE generation is controlled. Generation control of RE sources becomes essential in microgrids particularly when working in autonomous mode. When the network generation is less than the demand, then power balance can be maintained by availing energy storage support or by load shaping. However, when the situation reverses, and with the storage reaching its full capacity, then generation has to be reduced to maintain the network frequency. Therefore, the objective of generation control, on one hand, is to deliver available maximum power from the RE generators in order to maximize the RE utilization, if the same volume of power can be absorbed together by the loads and energy storage. On the other hand, when the demand on the network falls below the maximum possible generation, the operating point of the RE generators has to be shifted away from the maximum power point so as to maintain the power balance and stability.

2.4.8.1 MPPT of Solar PV

The power output of solar PV cell is non-linear with respect to solar irradiance, cell temperature and the load connected at its terminal, with a unique operating point at which the power output is at its maximum. All the three conditions change during the operation over a day as well as any time period. When it is intended to extract the maximum available power from the solar panel under all possible operating conditions, then the maximum power point tracking has to be incorporated. The typical i-v characteristics of a solar PV panel is shown in Figure 2.68, where the significant operating parameters are marked as V_{OC}, I_{SC}, P_{MPP}, V_{MPP} and I_{MPP}, representing the open circuit voltage, short circuit current, maximum power, voltage at maximum power and current at maximum power, respectively. The desired operating point

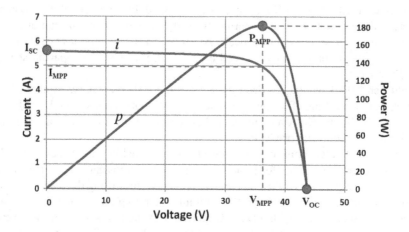

FIGURE 2.68 Typical i-v and p-v characteristics of a solar PV panel.

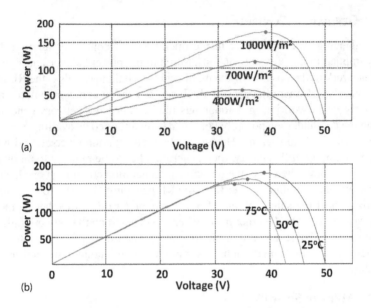

FIGURE 2.69 The *p-v* characteristics of PV cell with (a) varying irradiance and (b) varying temperature.

delivers the power of P_{MPP} at the corresponding voltage of V_{MPP} and current of I_{MPP}. Then it is necessary to adjust the effective resistance on the PV panel terminals so that the operation sustains at MPP.

When the irradiance and the temperature vary, then the maximum power point (MPP), described by P_{MPP}, V_{MPP} and I_{MPP}, also vary, as seen in Figure 2.69. The MPP excursions with changing environmental conditions necessitate inclusion of MPPT in PV systems. An MPPT algorithm if implemented will track MPP under any temperature and irradiance values.

Various methods for MPPT have been adopted in commercial PV systems, whereas their evolution is still continuing in the research arena aiming at performance enhancement in various aspects. Advanced MPPT techniques target to overcome additional non-linearity induced by non-uniform irradiance, partial shading conditions, etc.

Further, MPPT techniques diverge in computational complexity, number of sensors required, tracking efficiency, accuracy of MPP, convergence speed, implementation methods, etc. However, simple and easy-to-implement methods are the most preferred in commercial production.

Any MPPT algorithm can be tested for its tracking efficiency with EN 50530 European Standards for steady state as well as transient state performance. Most of these algorithms follow an iterative procedure in shifting the operation point of the PV panels until they reach the optimum operating point. Generally, a two-stage conversion is used with a DC-DC converter as the first stage and the inverter following it, before being fed either to grid or to load. Such schemes are illustrated in Figures 2.7, 2.8 and 2.15. The DC-DC stage connects the PV panel to the DC link and provides an instantaneous decoupling between the PV power generated and the power delivered to the grid. The choice of DC-DC converter depends on the obligatory matching between the PV array voltage and the required grid/load voltage. These DC-DC converters' duty ratio will be varied as per instructions by the MPPT algorithm. The MPPT algorithm, its dynamics and the DC link capacitance influence the quality of power fed to the grid/load.

The generic process followed in MPPT methods is to introduce a perturbation in an arbitrary direction in the DC link voltage and observe the power output of the PV array. If the power output treks up, then climb higher in the same direction until MPP is reached. Otherwise, if the power treks down upon a perturbation, then reverse the direction of perturbation and continue until P_{MPP} is reached. With a large value of DC link capacitance, a large volume of energy has to be transacted with the grid in every perturbation which causes considerable change in the DC link voltage. Also, such a large volume of power exchange will result in pulsations in the injected power. On the contrary, a small value of capacitance will propagate the PV power oscillations due to variations in temperature and irradiance to the grid. A MPPT algorithm needs to be evaluated, before choosing it in a solar PV system, based on several such characteristic features which influence the tracking performance.

In single stage grid-feeding solar PV system, where there is no DC-DC converter, the inverter itself is entrusted with the responsibility of keeping the operating point of the PV array at MPP. In such systems, MPPT will just select the right voltage at the PV panel output by control of power delivered by the inverter.

The most preferred MPPT methods, even in utility scale PV systems, are Perturb and Observe (P&O) and incremental conductance (IC).

- *P&O algorithm*: The P&O algorithm is implemented in the DC-DC converter stage by perturbation in its duty ratio, so as not to perturb the operating voltage of the PV array. If the response is an increase in power, the perturbation should be continued; else it should be reversed as described in Table 2.3

  Because of its iterative nature, the P&O method introduces oscillations around MPP, especially when working with rapidly changing environmental conditions. These oscillations can be minimized by reducing the perturbation size, but at the cost of a large convergence time for reaching

TABLE 2.3
The P&O Algorithm

Initial Direction of Perturbation	PV Power Ramp	Direction of Next Perturbation
Positive	Increase	Positive
Negative	Increase	Negative
Positive	Decrease	Negative
Negative	Decrease	Positive

MPP. Various improvisations are possible to overcome the limitations of the P&O method. One is adopting variable perturbation size; in other words, small size perturbation in the vicinity of MPP and large size otherwise. Estimating the initial operating point by use of AI techniques and subsequently introducing an appropriate magnitude and direction of perturbation, is another improvisation. Implementation of P&O algorithm needs at least two sensors, one for voltage and another for current, and can be done through microcontrollers.

- *Incremental conductance (IC) algorithm*: It is evident from the *p-v* characteristics that the slope of the curve is zero at MPP, while that on its left is positive and on the right is negative as shown in Figure 2.70.

Differentiating the power equation with respect to voltage, one can obtain the following classification:

$$P = VI$$

$$\frac{dP}{dV} = 0 \qquad\qquad (2.85)$$

$$\frac{dP}{dV} = I + V\frac{dI}{dV}$$

Therefore, at MPP, $-\frac{I}{V} = \frac{dI}{dV}$, on the left of MPP, $-\frac{I}{V} > \frac{dI}{dV}$, and on the right of MPP, $-\frac{I}{V} < \frac{dI}{dV}$.

It is possible at any instant of time to measure the panel current and voltage and calculate the instantaneous conductance, I/V. Subsequently, with a perturbation given to the duty ratio of the converter, if the current and voltage are measured at the next instant, the incremental conductance, $\Delta I/\Delta V$, also can be calculated. By comparing I/V and $\Delta I/\Delta V$ at every iteration it is possible to track the MPP. The trade-off between the iteration size and convergence time is a limitation when using the IC algorithm also.

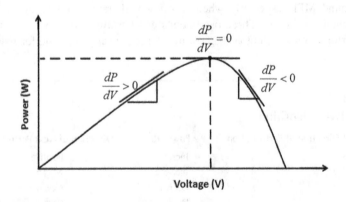

FIGURE 2.70 MPP tracking with IC method.

- *MPPT by DC-Link voltage control*: This method of MPPT is relevant and apt for the grid-feeding type of solar PV systems. The concept of this MPPT is explained in Figure 2.28 with the perspective of active power control in Section 2.4.1.3.1. Multistage solar PV systems operating in grid-feeding mode can be employed with this MPPT. Moreover, it requires neither measurements from PV panel nor computation of PV array power. Its implementation scheme is shown in Figure 2.71. If the inverter control loop is designed to maintain the DC-Link voltage constant, then the DC-DC converter can be made to work as a controlled current source delivering the current at MPP. With constant DC-Link voltage, $V_{dc\text{-}link}$, if the current command of the inverter control is increased continuously, then the power drawn from the DC-DC converter and the consequent power drawn from the PV panel can be increased. This can continue until the maximum available power from the panel is drawn and delivered to grid. The moment the current command requires an increase in the inverter power beyond P_{MPP}, then $V_{dc\text{-}link}$ starts falling. That means, the operation of the PV panel was at MPP just at the previous instant. This fall in $V_{dc\text{-}link}$ will cause the inverter controller to reduce the current command to return to MPP. The current command, I^*, is fed to the MPPT controller in order to vary the duty ratio and sustain it to keep the PV operating point at its optimum value. The method of calculation of I^* is the same as that described in Section 2.4.3.3.
- *Other MPPT algorithms*: There are several other MPPT algorithms ranging from iterative to non-iterative and using AI-based search methods to statistical data-based search algorithms, one cycle control, sliding mode control, state space model-based control, etc. The frequently used algorithms next to the aforesaid methods are briefly described below.

  Non-iterative MPPT methods: The non-iterative MPPT methods which hold short convergence time thanks to the absence of tracking. Fractional open circuit voltage (FOCV) method and fractional short circuit current (FSCC) method are popular in this category. Both of these methods follow the same concept to identify the MPP, but they use different control

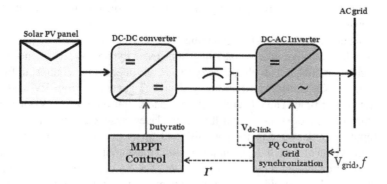

FIGURE 2.71 MPPT by DC-Link voltage control.

parameters. The FOCV method reaches MPP by establishing the panel voltage to be the identified V_{MPP}, while FSCC method establishes the panel current to be the identified I_{MPP}. Computation of V_{MPP} and I_{MPP} are possible because of the linear relationship between V_{MPP} & V_{OC} and I_{MPP} & I_{SC}, respectively and expressed as,

$$V_{MPP} = k_v V_{OC} \tag{2.86}$$

$$I_{MPP} = k_i I_{SC} \tag{2.87}$$

where k_v and k_i are the voltage and current constants and can be obtained from the PV panel data sheet. If V_{OC} or I_{SC} is measured periodically with the PV panel while in operation, then V_{MPP} or I_{MPP} can be calculated using equations 2.86 and 2.87, respectively. V_{MPP} and I_{MPP} serves as the reference for the converter control and the respective quantity is established at the panel output. However, implementation of these two methods of MPPT requires modification in the power circuit by way of additional hardware to open or shorten the panel circuit during the operation, yet with a reduction in the number of sensors to one. The temporary loss of power due to frequent OC and SC measurements is a major impediment here. Also, the constants in equations 2.86 and 2.87 will not be valid under partial shading conditions and will be influenced by aging – it challenges the accuracy of MPPT.

Fuzzy logic and neural network based MPPT: Working with vague inputs, with the ease of handling non-linear system dynamics and more importantly working even without an accurate mathematical model, the fuzzy logic controller has shown enormous potential to serve as an MPP tracker. As in any fuzzy logic control, the rule-based lookup table has to be developed for Fuzzy MPPT of any specific PV system. The accuracy and the convergence of Fuzzy MPPT depends on the effectiveness of the developed rule base. It is a set of rules which relate the errors and the control actions of the fuzzy controller. The inputs of the Fuzzy MPPT controller will generally be an error and/or the change in error. The error can be either in power, voltage or current, while the fuzzy logic controller output will typically be a change in duty ratio of the DC-DC converter.

Neural network (NN) is yet another MPPT tool which shares certain common features like ease in handling non-linear dynamics, lack of need for accurate model, etc., with the fuzzy logic controller. But in contrast to fuzzy control, NN control needs training data sets generated from PV systems tested over time. The NN architectures generally use three layers – namely, input layer, hidden layer, and the output layer arranged in the same order. Inputs like measured panel quantities and environmental data are presented to the system through the input layer, while the output layer delivers the duty cycle or change in duty cycle to the converter. So, the number of input and output nodes is decided by the number of inputs applied and the required number of outputs. But, the accuracy of MPPT depends on the number of hidden layers and the algorithm for the interconnection of the hidden layer nodes within as well as with the outer layers. The links connecting various nodes are assigned with weights and the identification of right weight for the desired input-output response is solely decided by the training process of NN. Since each PV system has its own unique *i-v* and *p-v* characteristics, and with even the response of the PV arrays being highly stochastic and site-dependent, frequent training would be necessary to obtain accurate MPPT operation; this is considered as a limitation of this method.

2.4.8.2 Off-MPP Operation of Solar PV

As already stated, the primary concern in the autonomous mode of operation of microgrids is to maintain the system stability through power balance control. When the demand on the network is high, then all RE generators are made to operate at the respective MPPs. On the other hand, it is essential during the lean load periods that the RE generators employ a power curtailment such that the generators are commanded to deliver the power just to meet the total demand. Such an off-MPP operation is possible by automatically moving the operating point to a suitable location off the MPP on the *p-v* curve of Figure 2.68. With respect to Figure 2.72, a solar PV system is operating initially at MPP with a load of P_{MPP} and if the system demand is falling down to P_{dem}, then, it is necessary to cut down the power generation through converter control. The necessity of power curtailment or off-MPP control can also be explained with the power versus time curve of a solar PV system shown in Figure 2.73. The power available in a typical day and the demand variation on a microgrid show that the amount of power to be curtailed varies over time and necessitates an additional control feature to follow this demand dynamics.

A simplified scheme of off-MPP generation control is presented in Figure 2.74. The required PV output power to meet P_{dem} is obtained as a current reference, I_{PV}^* and is compared with the measured I_{PV}. Finally, a current controller generates the required duty ratio to make I_{PV} follow its reference and accomplish the task of PV power curtailment.

The scheme suggested above is incomplete and needs to be worked out in detail to suit implementation in real applications. The strategy to estimate P_{dem}, the algorithm to switch between MPPT and off-MPP control modes, etc., can be developed only with knowledge of the specific topology of the microgrid and its desired functions.

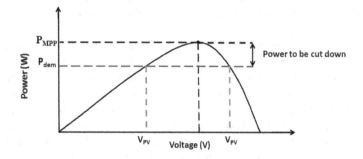

FIGURE 2.72   PV operation off MPP.

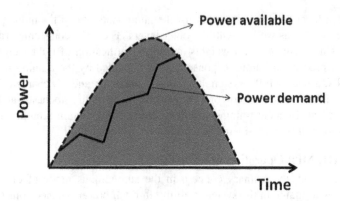

FIGURE 2.73 Variation in power demand and power available with respect to time.

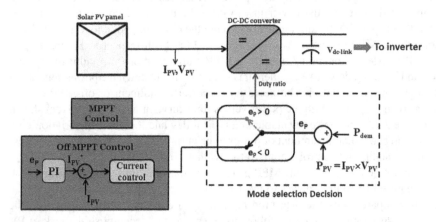

FIGURE 2.74 Off MPP control.

2.4.8.3 MPPT in Wind Turbine Generators

Maximum power point tracking algorithms employed in WT-Generator systems are discussed in Section 2.4.6.3. The WT is operated at MPP by maintaining the tip speed ratio, λ, of the WT at its optimum value, λ_{opt}, that corresponds to the maximum power coefficient, C_{pmax}.

The MPP in the characteristic curve of WT can be translated to power-speed characteristics of the WT driven generator (WTG) to help implementation of control. The locus of MPPs of the WTG for possible wind speeds, V_n, is depicted in Figure 2.75, wherein the dotted curves indicate the output power of WTG for each V_n with variation in the generator's rotor speed, ω_r. It can be noted that each curve has an MPP at which the electrical power output from the WTG is the maximum, P_{max}, for each V_n; the MPP locus is the line connecting all these P_{max} values. This MPP locus can be obtained through experiments, and it can be utilized to avail references for the development of MPPT in WTG systems.

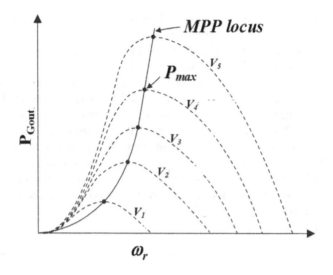

FIGURE 2.75 Locus of MPP of a WTG.

The MPPT controls of WT are implemented through power electronic converters; the control could be entrusted on different power converters depending on the type of generation system.

1. *MPPT in PMSG*: The MPPT control scheme for a WT driven grid synchronized PMSG system is shown in Figure 2.76. The MPPT control is similar to the schemes of Figure 2.60 described in the autonomous mode, while one of the schemes in Figures 2.33, 2.36, 2.49 or 2.52 can be used for grid-side converter control. The MPPT block in Figure 2.76 will use any of the MPPT techniques like TSR-MPPT, OT-MPPT or other advanced MPPT controls. It however generates the duty ratio for the MPPT converter. Once the MPPT converter ensures the operation of the WT at its λ_{opt}, then the

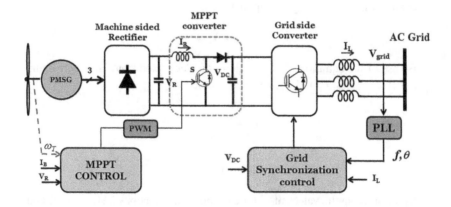

FIGURE 2.76 MPPT control of grid synchronized WT-PMSG system.

grid-side converter regulates the DC link voltage and pushes the maximum power into the grid at an appropriate amplitude of I_L.

2. *MPPT in DFIG*: The MPPT method for DFIG can be developed by obtaining the maximum power information through the locus of MPP curve. The output power, P_{Gout}, is the sum of stator output power and the rotor output power of DFIG. Depending on the operating mode like super- or sub-synchronous generation, the rotor power will be either positive or negative, respectively.

The MPPT can be implemented in DFIG through rotor current regulation using vector control techniques in stator flux oriented reference frame. The relationship between the rotor current and the stator power can be obtained from the fundamental equations of DFIG; subsequently, the reference current generation and the closed loop rotor current regulation can be accomplished.

The stator flux, ψ_s, of DFIG expressed in the stator flux-oriented reference frame is,

$$i_{ds}L_s + i_{dr}L_m = \psi_s \tag{2.88}$$

$$i_{qs}L_s + i_{qr}L_m = 0 \tag{2.89}$$

where i_{ds} and i_{qs} are the d- and q-axis stator currents, L_s is the stator inductance and L_m is the magnetizing inductance of DFIG. Using equations 2.88 and 2.89, it is possible to express the stator current components in terms of the rotor current components as,

$$i_{ds} = \frac{\psi_s}{L_s} - \frac{L_m}{L_s} i_{dr} \tag{2.90}$$

$$i_{qs} = -\frac{L_m}{L_s} i_{qr} \tag{2.91}$$

The stator flux is primarily set up due to the stator voltage established by the grid supply with the assumption that the stator resistance drop is negligibly small. Therefore, with the stator angular frequency, ω_s, it is possible to express the stator voltage components as,

$$v_{ds} = 0 \; ; \; v_{qs} = V_g = \omega_s \psi_s \tag{2.92}$$

By use of the equations 2.37 and 2.38, the decoupled stator side active and reactive powers, P_s and Q_s, can be expressed as,

$$P_s = \frac{3}{2} v_{qs} i_{qs} \tag{2.93}$$

$$Q_s = -\frac{3}{2} v_{qs} i_{ds} \tag{2.94}$$

By substituting the values of stator currents from equations 2.90 and 2.91 and the stator voltage from equation 2.92 in the power equations of 2.93 and 2.94,

$$P_s = -\frac{3}{2} v_g \frac{L_m}{L_s} i_{qr} \tag{2.95}$$

$$Q_s = \frac{3}{2} \frac{v_g^2}{\omega_s L_s} - \frac{3}{2} v_g \frac{L_m}{L_s} i_{dr} \tag{2.96}$$

The reference currents to be delivered to the rotor to accomplish the stator side powers of P_s and Q_s can be calculated from equations 2.95 and 2.96.

The scheme for realizing MPPT in DFIG is presented in Figure 2.77. The DFIG rotor speed, ω_r, is measured and it is projected onto the locus of MPP to find the corresponding P_{max}, which will serve as the power reference, P_s^*, for the rotor side converter control along with a Q_s^*. The active and reactive power flows on the stator can be controlled through regulation of d- and q-axes components of the rotor current. The power references are converted as rotor current references using equations 2.95 and 2.96. Further, these references are converted in the stator flux orientated reference frame with the help of the rotor flux angle, θ_r. As seen in Figure 2.77, θ_r is obtained from the measured rotor speed, ω_r, and the stator flux vector, θ_s, of the grid voltage. Further, the PI compensators in the current control loop ensure that the actual currents follow the rotor current references.

2.4.8.4 Off-MPP Control in WTG

The Off-MPP control of WTG is necessary during autonomous mode of operation of the microgrid as in the case of PV systems too. Such power curtailment control ensures power balancing in AC network as well as voltage and frequency regulation. Figure 2.78 gives the C_p-λ characteristics of WT; its features have already been discussed. In Figure 2.78, MPP is represented by the point, $(\lambda_{opt}, C_{pmax})$, whereas off-MPP operating points are represented by (λ_1, C_{p1}), (λ_2, C_{p2}), etc. The tip speed ratio being defined as,

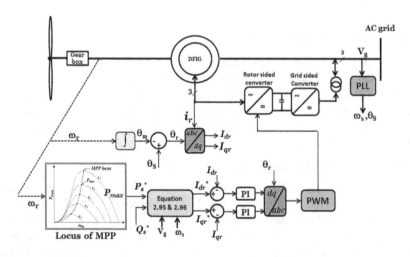

FIGURE 2.77 MPPT control scheme for DFIG.

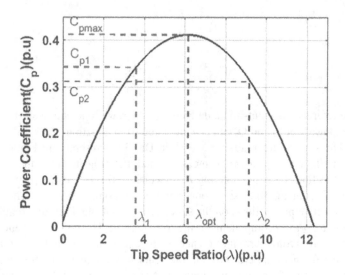

FIGURE 2.78 C_p-λ characteristics of WT.

$$\lambda = \frac{R\omega_T}{v_W} \tag{2.97}$$

where ω_T is WT shaft speed, R is turbine blade length and v_w is the wind speed, and the shaft speeds of WT and DFIG being rigidly related by the gear ratio, λ can be projected as ω_r.

By keeping the load demand, P_{dem}, as the reference to the control loop, the off-MPP algorithm will identify the value of the generator shaft speed at which the required power can be generated at a reduced C_p. Further, the same procedure is followed in the control loop of Figure 2.77 to make DFIG deliver the power demand identified with the reduced C_p.

2.4.9 CONTROL FOR POWER QUALITY

Power quality in microgrid is of much concern due to the presence of power electronic based generation units and intermittent RE based generators feeding large number of non-linear loads as well as single phase loads. This becomes crucial especially in the autonomous mode of operation. The power quality can be referred as the capacity of the microgrid system to deliver and distribute pure sine power at rated voltage and rated frequency.

2.4.9.1 Power Quality Standards

The power quality issues are quantified and categorized based on various standards like IEEE 519-1992, IEC 61000-4-30, and EN50160 across the globe. The power quality in microgrid is influenced by two factors: (i) Voltage quality and (ii) continuity of power supply. The former includes voltage fluctuations, voltage and phase unbalance and harmonics in voltage waveforms, while the latter includes the type and duration of interruptions, voltage levels during faults, etc.

Any distortion in the voltage or current represents the presence of frequency components other the fundamental frequency. Fundamental frequency component is responsible for active power delivery, while currents at other frequencies account for harmonic power losses and associated implications on system voltage distortion. The load current harmonics passing through the system impedance will cause non-sinusoidal voltage drop, which in turn will distort the load voltage at common connection point. So, measures should be taken to restrict the current harmonic magnitudes in microgrids which are caused by (i) non-linear loads and (ii) the harmonic rich currents delivered by the RE converters.

Harmonic distortions are quantified by total harmonic distortion (THD), which is a measure of the number of harmonics with respect to the fundamental frequency component in a given waveform. The THD of voltage and THD of current are expressed as,

$$\text{THD}_I = \frac{\sqrt{\sum_{n=2,3,4,5,\dots}^{\infty} I_n^2}}{I_{1rms}} = \frac{\sqrt{I_{rms}^2 - I_{1rms}^2}}{I_{1rms}} \tag{2.98}$$

$$\text{THD}_V = \frac{\sqrt{\sum_{n=2,3,4,5,\dots}^{\infty} V_n^2}}{V_{1rms}} = \frac{\sqrt{V_{rms}^2 - V_{1rms}^2}}{V_{1rms}} \tag{2.99}$$

where n is the harmonic order, I_{rms} is the rms value of the distorted current waveform, and I_{1rms} is the rms value of the fundamental frequency component of the current waveform, V_{rms} is the rms value of the distorted voltage waveform, and V_{1rms} is the rms value of the fundamental frequency component of the voltage waveform. Ideally, the THD of a waveform is expected to be zero while the permissible limits of THD, as prescribed by IEEE 519, are presented in Tables 2.4 and 2.5.

The THD quantifies the current distortion levels, but this can be a pessimistic value unless it is correlated with the load current magnitude. Large THD with low magnitude of current may not pose a potential threat on the voltage at PCC. So, the denominator of THD calculation is revisited as the fundamental of the peak demand load current rather than the fundamental of the present load current to form a new index called total demand distortion (TDD). The IEEE 519 defines TDD as

TABLE 2.4
Voltage Distortion Limits

Bus Voltage at PCC	Individual Voltage Distortion (%)	Total Voltage Distortion THD (%)
69 kV and below	3.0	5.0
69.001 kV through 161 kV	1.5	2.5
161.001 kV and above	1.0	1.5

TABLE 2.5

Current Distortion Limits for General Distribution Systems (120 V–69 kV)
h: individual harmonic order (Odd Harmonics)

I_{sc}/I_L	<11	11≤h<17	17≤h<23	23≤h<35	35≤h	TDD
<20[a]	4.0	2.0	1.5	0.6	0.3	5.0
20 < 50	7.0	3.5	2.5	1.0	0.5	8.0
50 < 100	10.0	4.5	4.0	1.5	0.7	12.0
100 < 1000	12.0	5.5	5.0	2.0	1.0	15.0
>1000	15.0	7.0	6.0	2.5	1.4	20.0

Note: Even harmonics are limited to 25% of the odd harmonic limits given above.

[a] All power generation equipment is limited to these values of current distortion, regardless of actual I_{sc}/I_L, where I_{sc} = Maximum short circuit current at PCC, I_L = Maximum demand fundamental frequency load current at PCC.

"The total root-sum-square harmonic current distortion, in percent of the maximum demand load current," and expressed as,

$$\text{TDD} = \frac{\sqrt{\sum_{n=2,3,4,5,...}^{\infty} I_n^2}}{I_L} = \frac{\sqrt{I_{rms}^2 - I_{1rms}^2}}{I_L} \tag{2.100}$$

where I_L is the fundamental component of the maximum demand load current at PCC when monitored typically between 15 and 30 minutes as per IEEE519.

The IEEE519 further defines the allowed percentage of individual harmonics at PCC as a function of the load magnitude. This magnitude is defined as short-circuit ratio, which is the ratio of the maximum short-circuit current, I_{SC}, at PCC to the maximum value of the fundamental load demand current, I_L, at PCC as described in Table 2.5.

2.4.9.2 Power Quality Control

The primary power quality control should be initiated in the RE converters, whose control and the output filter have to be designed to minimize the harmonic current injected into the microgrid. These have to meet the harmonic standards as specified in Table 2.5. However, the load induced harmonics can be mitigated through filters. Shunt and series filters are designed and used at appropriate locations of the microgrid to reduce current harmonics. Further, tuned harmonic filters are used to selectively eliminate dominant harmonics depending on the type of non-linear loads. But, the cost of the passive filters become prohibitively high at high power levels which makes active power filters a natural choice. An active filter is basically a voltage source inverter controlled by digital processors. It generates a compensating current in tune with the load current such that the current drawn from the AC grid is pure sine. The concept of an active filter is presented in the block diagram of

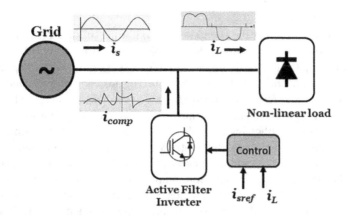

FIGURE 2.79 Concept of active filter.

Figure 2.79, wherein the source current, i_s, can be controlled to remain sinusoidal by continuously sensing the load current, i_L, and accordingly switching the inverter.

Another predominant power quality issue is voltage imbalance due to uneven distribution of single-phase loads within the microgrid and in particular the large single-phase loads. Such unbalanced voltages will cause derating of three phase equipment like induction motors, failure of rectifier-fed loads, etc. Series and shunt filters are among the recommended solutions for voltage imbalance; yet another common method is negative sequence voltage/current injection.

Nevertheless, the type of power quality control in microgrids needs to be selected based on the type of loads, the type of distortions, the topology of the microgrid and importantly the project economics. Advanced devices like unified power quality controller (UPQC), unified power flow controller (UPFC), etc., are some of the new generation power quality correction methods. These are versatile in correcting multiple power quality issues and can be included in the microgrid as future power quality devices.

2.4.10 ISLANDING DETECTION AND PROTECTION SCHEMES

Islanding is a condition in an electrical network in which one or more of generating sources and loads get isolated from the main grid or utility. Islanding happens when the grid disconnects at the PCC of a microgrid and the local generators of the latter continue to supply power to the local loads and form an electrical island. The grid disconnection may happen due to one or a combination of the reasons described below:

- A grid fault identified by the utility protection devices that initiated a disconnection,
- An equipment failure initiated opening of grid supply,
- An intentional shut down of the utility for network maintenance, or
- Natural calamities disrupted the network.

2.4.10.1 Need for Islanding

Island condition can affect the normal operation of the utility as it is a hazardous mode of operation and can cause safety problems for utility personal, if live generators on the load side remain connected to distribution network in spite of grid failure and disconnection. On the contrary, microgrids can opt for islanding, with proper isolation for safety, for several favorable reasons like increased reliability for local loads, higher system efficiency, high power quality and tariff benefits. So, islanding can be either intentional, like the case of microgrids opting for it, or unintentional due to reasons on the utility side. In any case, an islanding detection and isolation unit is essential in every microgrid which has the responsibility to detect the absence of mains and communication of that information to various control nodes, as in Figure 3.33 of Chapter 3. Real-time information on islanding is essential because several microgrid resources are designed to work in multiple modes like grid-forming, grid-feeding, grid-support, MPPT, off-MPP, etc., where the control strategies are based on whether the microgrid is in grid-connected mode or autonomous mode.

Detection of unintentional islanding is significant and also challenging and research has contributed much in this direction.

2.4.10.2 Islanding Detection Methods – Passive and Active Schemes

Islanding detection (ID) methods are classified into four broad types – namely, passive methods resident in inverters, active methods resident in inverters, remote methods or communication-based methods resident or non-resident in inverters, and active methods non-resident in inverters. The last method is not discussed here as it is rarely used in any application.

Figure 2.80 shows the equivalent circuit of a grid-connected inverter feeding power to grid. The local loads are modeled as a parallel RLC circuit and the equivalent grid impedance is modeled as a series RL components.

Generally, RLC loads show up a high value of quality factor, $Q = R\sqrt{\frac{C}{L}}$ with large value of C and small values of L and R. Such a high value of Q represents the amount of energy exchanged and dissipated in the loads, which leads to large non-detection zones (NDZ) in the islanding detection process. The NDZ is an indication of failure in islanding detection.

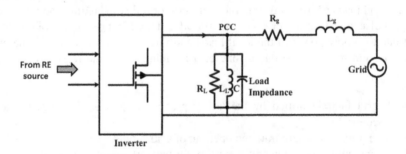

FIGURE 2.80 Equivalent circuit of inverter feeding the grid.

A PV distributed power generation system must comply with IEEE 1547-2018 or IEC 62116 for a Q factor of 1 and IEEE 929 for a Q factor of 2.5. These standards accommodate distributed energy resource challenges like intentional islanding and impact on local area network. All these standards suggest the same maximum detection time of 2 s, with the allowed frequency variation of about ± 1.5 Hz (IEC), -0.7 to $+0.5$ Hz (IEEE) and with an allowed voltage variation of $\pm 15\%$ (IEC) or -12% to $+10\%$ (IEEE).

2.4.10.2.1   Passive ID Methods

Passive ID methods observe the electrical quantities of the grid – such as voltage, current, frequency, etc., through continuous measurement and monitoring; any deviation from the nominal value is an indication of absence of the mains. Passive ID methods are characterized by large NDZ especially when working with large Q factor loads. Major types of passive ID methods are:

- Over/under voltage protection (OVP/UVP),
- Over/under frequency protection (OFP/UFP),
- Rate of change of frequency (ROCOF),
- Voltage phase jump, and
- Detection of voltage/current harmonics.

The passive ID methods like over/under voltage and over/under frequency utilize the inherent over/under frequency protections and over/under voltage protections available in the grid-connected inverters. These are the mandatory protections which will cause the inverters to trip the circuit and stop supplying power to grid at the event of frequency or voltage at PCC cross the prescribed limit. The protection system data is used for islanding detection too. However, these methods exhibit relatively large NDZ with unpredictable detection time.

ROCOF based ID computes the rate of change of frequency, df_{grid}/dt in a short time window. During the operation of microgrids in grid-connected mode, just prior to the failure of main grid, power balance exists at PCC; at the instant of grid failure a power imbalance will be created momentarily which will tend to deviate the frequency at PCC. If this change in frequency is measured and compared with a predefined threshold value, it can identify the islanding condition. The ROCOF computation and their time windows and however, introduce large NDZ.

Phase jump detection method is based on the measurement of phase difference between the inverter's voltage and the respective current.

When the grid fails, the phase angle at which current is delivered by the inverter is solely decided by the local loads, while it was earlier decided by the reactive power command. A sudden jump in the phase difference indicates islanding. The changed phase angle is compared with a threshold value to detect occurrence of islanding. But choosing right threshold value of phase angle is challenging as any change in the load phase angle should not detect a false islanding condition. A PLL can be employed to detect the phase angle in this method.

All grid-connected inverters are provided with the feature to calculate the voltage and current THDs. When the grid is present, the inverter current drops across very small grid impedance and so the voltage THD will comply with the standards. But when the grid fails, the inverter current entirely flows into the local loads whose impedance will be much higher than the grid impedance and it causes a high value of voltage THD. This change in THD can be utilized to detect islanding condition.

2.4.10.2.2   Active ID Methods

Active methods introduce a perturbation at the output of the RE generators at PCC by injecting a signal, which causes the system parameters to deviate from their nominal values under islanded conditions. Care needs to be taken to avoid power quality problems caused by the injected signals in the electric utility. Most of the active ID methods have very small NDZ.

Some of the frequently used active ID methods are described as follows:

- *Impedance measurement*: This method is based on the change in impedance, looking from the PCC, while sliding from grid-connected to islanded mode. Injection of a high-frequency signal through grid-feeding inverter and subsequent voltage and current measurements can be used to calculate the impedance. The impedance of local loads being larger than the grid impedance, the change in impedance is an indication of the presence/absence of the grid.
- *Active frequency drift*: The waveform of the current injected by the inverter into grid is introduced with a distortion such that a continuous drift in the frequency will happen. The frequency will be unaffected in grid-connected mode whereas frequency at PCC will drift up or down, depending on the introduced distortion, when islanding occurs.
- *Sandia frequency shift*: This method is similar to the active frequency shift, but with an additional positive feedback. The methodology remains the same as far as the injection and behavior under grid-tied connection. But when the grid fails, a positive feedback increases the frequency error which further increases the inverter output frequency and elevates the frequency drift to higher values, which results in reduced NDZ.

Other active ID methods include Sandia voltage shift, slide mode frequency shift, automatic phase shift, adaptive logic phase shift, frequency bias, frequency jump, proportional power spectral density, detection of impedance at a specific frequency, etc.

2.4.10.2.3   Communication Based ID Methods

These methods can be resident or non-resident in the inverter, but they are implemented through the communication between the inverter and the grid in order to trip the inverter when the islanding condition is detected.

Power Line Carrier Communications (*PLCC*): The PLCC is a traditional communication system used in power systems. More intricate details of PLCC are dealt in Section 3.2.1.3 of Chapter 3. It sends signals of low-energy and high frequency which are modulated and superimposed on the power signals in the power line. By locating a receiver on the microgrid-connected to such PLCC enabled grids, it is possible to detect the loss of PLCC signals. This loss of communication signal indicates the islanding condition. Also, the PLCC receiver can be made to communicate to the inverters about the islanding and take further actions in autonomous mode.

Supervisory Control and Data Acquisition (SCADA): The SCADA is used for continuous monitoring and automation of power system equipment like circuit breakers, generators, transformers or the substation as a whole. The SCADA has sensors to monitor electrical quantities like voltage, current, frequency, power, etc., and can detect loss of any of these quantities in the system. When islanding occurs, substantial deviation in the grid parameters will also happen, which can be treated as indicators for ID.

The second decade of the twenty-first century witnessed the merits of signal processing techniques when fused into several passive ID methods to accomplish superior detection performance with small NDZ, short detection time and enhanced dynamic response. The frequency domain and time-frequency correlated signal processing techniques like fast Fourier transform, wavelet transform, Kalman filter, empirical mode decomposition, empirical wavelet transforms and variational mode decomposition have been attempted by several researchers. These techniques offer feature extraction of grid signals, signal decomposition, and frequency classification to distinguish between the normal mode of operation and islanded mode.

BIBLIOGRAPHY

1. M. A. Hossain, H. R. Pota, M. J. Hossain and F. Blaabjerg, "Evolution of microgrids with converter-interfaced generations: Challenges and opportunities," *International Journal of Electrical Power & Energy Systems*, vol. 109, pp. 160–186, 2019.
2. S. Muller, M. Deicke and R. W. De Doncker, "Doubly fed induction generator systems for wind turbines," *IEEE Industry Applications Magazine*, vol. 8, no. 3, pp. 26–33, 2002.
3. B. Arbab-Zavar, E. Palacios-Garcia, J. Vasquez and J. Guerrero, "Smart inverters for microgrid applications: A review," *Energies*, vol. 12, no. 5. p. 840, 2019. doi:10.3390/en12050840.
4. H. Pourbabak, H. P. Pourbabak, T. C. Chen, B. Z. Zhang and W. S. Su, "Control and energy management system in microgrids," *Clean Energy Microgrids*, pp. 109–133. doi:10.1049/pbpo090e_ch3.
5. S. Tahir, J. Wang, M. Baloch and G. Kaloi, "Digital control techniques based on voltage source inverters in renewable energy applications: A review," *Electronics*, vol. 7, no. 2. p. 18, 2018. doi:10.3390/electronics7020018.
6. J.-W. Chang, G.-S. Lee, H.-J. Moon, M. B. Glick and S.-I. Moon, "Coordinated frequency and state-of-charge control with multi-battery energy storage systems and diesel generators in an isolated microgrid," *Energies*, vol. 12, no. 9. p. 1614, 2019. doi:10.3390/en12091614.
7. S. Mishra, "Permanent magnet synchronous generator based stand-alone wind energy supply system," *2013 IEEE Power & Energy Society General Meeting*. 2013. doi:10.1109/pesmg.2013.6673039.
8. M. E. Haque, K. M. Muttaqi and M. Negnevitsky, "Control of a stand alone variable speed wind turbine with a permanent magnet synchronous generator," *2008 IEEE Power and Energy Society General Meeting–Conversion and Delivery of Electrical Energy in the 21st Century*, 2008. doi:10.1109/pes.2008.4596245.

9. N. A. Orlando, M. Liserre, R. A. Mastromauro and A. Dell'Aquila, "A survey of control issues in PMSG-based small wind-turbine systems," *IEEE Transactions on Industrial Informatics*, vol. 9, no. 3. pp. 1211–1221, 2013. doi:10.1109/tii.2013.2272888.

10. O. Palizban, K. Kauhaniemi and J. M. Guerrero, "Microgrids in active network management – part II: System operation, power quality and protection," *Renewable and Sustainable Energy Reviews*, vol. 36. pp. 440–451, 2014. doi:10.1016/j.rser.2014.04.048.

11. F. Ewald and M. A. S. Masoum, *"Power Quality in Power Systems and Electrical Machines."* Academic Press, 2011.

12. A. F. Zobaa and S. H. E. A. Aleem, *"Power Quality in Future Electrical Power Systems."* Institution of Engineering & Technology, 2017.

13. A. Kusko and M. T. Thompson, *"Power Quality in Electrical Systems."* vol. 23, New York: McGraw-Hill, 2007.

14. Y. Yang, F. Blaabjerg and H. Wang, "Constant power generation of photovoltaic systems considering the distributed grid capacity," *2014 IEEE Applied Power Electronics Conference and Exposition – APEC 2014*. 2014. doi:10.1109/apec.2014.6803336.

15. Thongam J.S. and M. Ouhrouche, MPPT control methods in wind energy conversion systems, *Fundamental and Advanced Topics in Wind Power*, no.1, pp. 339–360, 2011. doi:10.5772/21657.

16. S. P. Manoj, A. Vijayakumari and S. K. Kottayil, "Development of a comprehensive MPPT for grid-connected wind turbine driven PMSG," *Wind Energy*, vol. 22, no. 6. pp. 732–744, 2019. doi:10.1002/we.2318.

17. A. Vijayakumari, A. T. Devarajan and N. Devarajan, "Decoupled control of grid-connected inverter with dynamic online grid impedance measurements for micro grid applications," *International Journal of Electrical Power & Energy Systems*, vol. 68. pp. 1–14, 2015. doi:10.1016/j.ijepes.2014.12.015.

18. R. Teodorescu, M. Liserre and P. Rodriguez. *"Grid Converters for Photovoltaic and Wind Power Systems."* vol. 29. John Wiley & Sons, 2011. doi:10.1002/9780470667057.

19. W. Bower and M. Ropp, Evaluation of islanding detection methods for utility-interactive inverters in photovoltaic systems, *Sandia Report SAND*, vol. 3591, p. 2002.

20. M.-S. Kim, R. Haider, G.-J. Cho, C.-H. Kim, C.-Y. Won and J.-S. Chai, "Comprehensive review of islanding detection methods for distributed generation systems," *Energies*, vol. 12, no. 5. p. 837, 2019. doi:10.3390/en12050837.

21. S. Whaite, B. Grainger and A. Kwasinski, "Power quality in DC power distribution systems and microgrids," *Energies*, vol. 8, no. 5. pp. 4378–4399, 2015. doi:10.3390/en8054378.

22. F. Nejabatkhah, Y. W. Li and H. Tian, "Power quality control of smart hybrid AC/DC microgrids: An overview," *IEEE Access*, vol. 7. pp. 52295–52318, 2019. doi:10.1109/access.2019.2912376.

23. Y. Liu et al., "A reactive power-voltage control strategy of an AC microgrid based on adaptive virtual impedance," *Energies*, vol. 12, no. 16. p. 3057, 2019. doi:10.3390/en12163057.

24. S. Tahir, J. Wang, M. Baloch and G. Kaloi, "Digital control techniques based on voltage source inverters in renewable energy applications: ArReview," *Electronics*, vol. 7, no. 2. p. 18, 2018. doi:10.3390/electronics7020018.

25. T. Strasser et al., "A review of architectures and concepts for intelligence in future electric energy systems," *IEEE Transactions on Industrial Electronics*, vol. 62, no. 4. pp. 2424–2438, 2015. doi:10.1109/tie.2014.2361486.

26. A. Gkountaras, *"Modeling Techniques and Control Strategies for Inverter Dominated Microgrids."* Vol. 2, Berlin: Universitätsverlag der TU, 2017.

27. J. Rocabert, A. Luna, F. Blaabjerg, and P. Rodríguez, "Control of power converters in AC microgrids," *IEEE Transactions on Power Electronics*, vol. 27, no. 11. pp. 4734–4749, 2012. doi:10.1109/tpel.2012.2199334.

28. B. Kroposki, B. Johnson, Y. Zhang, V. Gevorgian, P. Denholm, B. M. Hodge and B. Hannegan, "Achieving a 100% renewable grid: Operating electric power systems with extremely high levels of variable renewable energy," *IEEE Power and Energy Magazine*, vol. 15, no. 2, pp. 61–73, 2017.

29. D. Kumarja and K. Chatterjee, "A review of conventional and advanced MPPT algorithms for wind energy systems," *Renewable and Sustainable Energy Reviews*, vol. 55, pp. 957–970, 2016. doi:10.1016/j.rser.2015.11.013.

30. B. Arbab-Zavar, E. Palacios-Garcia, J. Vasquez and J. Guerrero, "Smart inverters for microgrid applications: A review," *Energies*, vol. 12, no. 5. p. 840, 2019. doi:10.3390/en12050840.

31. M.-S. Kim, R. Haider, G.-J. Cho, C.-H. Kim, C.-Y. Won and J.-S. Chai, "Comprehensive review of islanding detection methods for distributed generation systems," *Energies*, vol. 12, no. 5. p. 837, 2019. doi:10.3390/en12050837.

32. Z. Ali, N. Christofides, L. Hadjidemetriou, E. Kyriakides, Y. Yang and F. Blaabjerg, "Three-phase phase-locked loop synchronization algorithms for grid-connected renewable energy systems: A review," *Renewable and Sustainable Energy Reviews*, vol. 90. pp. 434–452, 2018. doi:10.1016/j.rser.2018.03.086.

33. Y. Naderi et al., "Power quality issues of smart microgrids: Applied techniques and decision making analysis," *Decision Making Applications in Modern Power Systems*, pp. 89–119, 2020. doi:10.1016/b978-0-12-816445-7.00004-9.

34. M.-S. Debry, G. Denis and T. Prevost, "Characterization of the grid-forming function of a power source based on its external frequency smoothing capability," *2019 IEEE Milan Power Technology*, 2019. doi:10.1109/ptc.2019.8810409.

35. K. W. Joung, T. Kim and J.-W. Park, "Decoupled frequency and voltage control for stand-alone microgrid with high renewable penetration," *IEEE Transactions on Industry Applications*, vol. 55, no. 1. pp. 122–133, 2019. doi:10.1109/tia.2018.2866262.

36. H. D. Tafti, A. I. Maswood, G. Konstantinou, J. Pou and P. Acuna, "Active/reactive power control of photovoltaic grid-tied inverters with peak current limitation and zero active power oscillation during unbalanced voltage sags," *IET Power Electronics*, vol. 11, no. 6. pp. 1066–1073, 2018. doi:10.1049/iet-pel.2017.0210.

37. U. B. Tayab, M. A. B. Roslan, L. J. Hwai and M. Kashif, "A review of droop control techniques for microgrid," *Renewable and Sustainable Energy Reviews*, vol. 76. pp. 717–727, 2017. doi:10.1016/j.rser.2017.03.028.

3 Communication Infrastructure for Smart Microgrids

P. Sivraj

CONTENTS

3.1 THE NEED TO TURN "SMART"

All major innovations paving the way for a smarter world were either directly or indirectly connected to developments in the industrial sector. When we look at industrial revolutions, historically, they address major technological advancements in the field of manufacturing and associated industries. With each incremental phase of industrial revolution, the changes and modernization became more and more cyber and less physical leading to cyber-physical systems where the cyber system is an essential feature that adds smartness to all physical systems and services. As a result, transformations today are more on the digital side in industries, treating production and operation as services of smart systems revealing new business models and solutions. Such transformations bringing in smart systems are not limited to industrial systems but extend to almost all application domains in the real world. The transformation of various domains and services into smart systems is more of a natural evolutionary process than a forced integration of smartness into these systems (Figure 3.1).

A power system network is an infrastructural system designed, setup and maintained for delivery of electrical energy from electricity generating stations at different locations to different types of consumers who are geographically distributed. A close understanding of the power system evolution over the last century will reveal the slow and steady way in which *smartness* has crept into the various sectors of the network. The need to shift to smart grids has mainly been two-fold – namely, natural and progressive evolution and pressing demand for improved and innovative services from all stakeholders. All bulk generation plants are being automated with SCADA systems reducing the need for human intervention in managing the operation of generation plants. This automation is also expected to fulfill the automated control of generation and grid integration

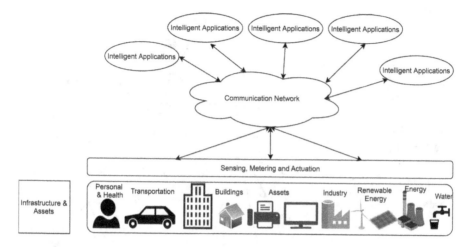

FIGURE 3.1 Smart services leading to smart infrastructure.

based on demand and grid status. The transmission sector has substations distributed at different geographic locations and overhead transmission lines to transfer electrical energy from one location to another. The modernization and automation in the transmission sector involves automating the operation of these substations and also enabling connectivity among these substations and control centers. Another aspect of automation is the real-time monitoring of the power transmission network for optimal control and protection. The same needs of substation automation and grid monitoring, control and protection are essential in the power distribution network, also.

Generation, which used to be primarily centralized and bulk, has been transformed into a mixture of centralized and distributed, with options for distributed energy storage, too. Also, the emphasis and focus of the new technologies are on the development of more distributed generation plants integrating electrical energy from renewable resources, thereby helping to realize the concept of microgrids. With the advent of technology that can enable fruitful and seamless integration of distributed generation with the grid, *con*sumers are becoming *pro*sumers by having captive RE power plants on one hand and exploring import-export transactions with the grid on the other hand.

The metamorphosis of the electric power system is also being shaped by the new operational features and management structure of the evolving hardware infrastructure. Whereas in the past, power flow was unidirectional and power system operation was monopolistic, the bidirectional power flow in the networked lines as well as participatory operation of the business has necessitated active communication in the system, among the nodes of the network and between the utility, operator and con(pro)sumer. Such communication involves data exchange, handshakes, switching commands, permissions, bills, bids, etc (Figure 3.2).

All these changes clearly bring out real-time data collection, processing, message passage and decision-making capabilities as unavoidable requirements to realize the

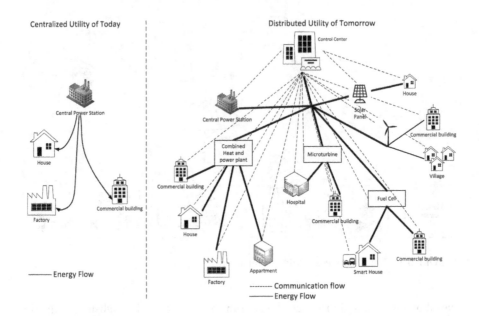

FIGURE 3.2 Traditional power grid vs smart grid.

vision of a smart grid. More than a simple networked system the communication network added to power system is expected to bring in required intelligence at all levels of operation. The smartness envisaged for the future grid is expected to convert the conventional power delivery system to a reliable, uninterrupted, affordable and universal system.

3.2 COMMUNICATION TECHNOLOGIES AND STANDARDS

Communication technology or information and communication technologies (ICT) refers to all software and hardware modules that are used in a system to communicate information or exchange data. A communication network is a collection of nodes or network elements that can interact with each other for information exchange over a shared resource such as wired or wireless medium. Realization of a network for data exchange in any application domain happens only with the help of communication technologies. Various network elements can be interconnected using the same communication technology to form a homogeneous communication network or by different communication technologies to form a heterogeneous communication network depending on the communication requirements of the applications. Depending on the number of member nodes, area covered, amount and rate of data transfer, the network can be classified typically into local area network (LAN), metropolitan area network (MAN) and wide area network (WAN). The ICT is an umbrella term that includes communication over any medium and the various services associated with it (Figure 3.3).

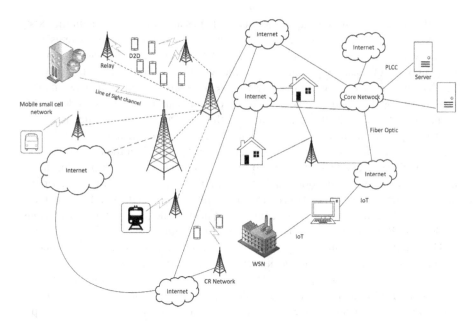

FIGURE 3.3   Smart infrastructure with heterogeneous communication.

There exist a large variety of wired and wireless communication technologies that can be used for interconnecting various elements of the smart power system network to create an ICT network for message exchange. A few technologies that fit the purpose are listed as follows:

1. Wired
   a. Fiber optic communication
   b. Ethernet
   c. Power line carrier communication
2. Wireless
   a. GPRS/LTE
   b. Wi-Fi
   c. WiMAX
   d. ZigBee
   e. Bluetooth
   f. Cognitive radio
   g. Wavenis
   h. HomePlug
   i. 6LoWPAN
   j. Z-Wave
   k. Wireless HART
   l. Insteon
   m. Wireless M-Bus

With the advent of new communication technologies and improvements in the existing ones, ICT is turning out to be the backbone of automation in power systems, paving the way to the realization of the smart grid. As power system automation demands large amounts of message passing among the widely distributed nodes in real time, the challenge involved in making a power system "smart" is manifold. The data that is generated by distributed devices must be communicated to multiple nodes at different hierarchical levels of operation to meet the large spectrum of applications with diverse requirements in the new grid. The choice of communication technology is not easy, as there are many technologies that can potentially suit the requirements of a given application and therefore the optimized selection needs a careful performance analysis. As the communication requirements of applications across different sectors and hierarchical levels of smart grid are largely diverse, the end-to-end communication architecture in a smart grid system will be a heterogeneous one. It poses a serious challenge to ICT integration within power systems (Figure 3.4).

This challenge is addressed by means of *standards*. The standards of a particular communication technology define its features, including performance and operational aspects, in establishing interconnection of multiple devices to form a network. The standards of various smart power system applications define the respective communication requirements for message exchange in terms of message sequence, type, rate, participating members and other aspects. A few of the many available standards are listed as follows:

1. IEEE 1646-2004
2. IEEE 1547.3-2007
3. IEC 61850-90-5
4. IEEE C37.118.2
5. IEEE 1815-2012
6. DLMS/COSEM – IEC 62056
7. IEEE 2030

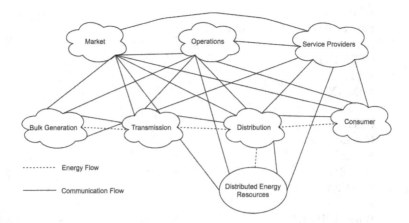

FIGURE 3.4   ICT interconnecting various stakeholders of power grid (based on NIST model).

3.2.1 WIRED COMMUNICATION TECHNOLOGIES

Wired communication technologies use a wired physical medium for exchange of data from one network element to another. Use of hardwire connection provides better security and minimal interference along with additional benefits like large bandwidth, better data rate and range. High installation and maintenance cost, right of way issues to lay the cables and network scalability are some of the disadvantages of wired medium. Some examples of typical wired communication systems are telephone, ethernet, cable television network, and fiber optic communication. There are various standards and protocols that govern the data communication over wired medium. These protocols and standards define the operational measures and specifications of medium for data transmission among multiple nodes. Most commonly used media for wired communication include coaxial cable, twisted pair copper cable, fiber optic cable and power line.

3.2.1.1 Ethernet – IEEE 802.3

Ethernet represents a variety of data communication techniques over wired medium commonly used for computer network communication in LAN, MAN and WAN. Ethernet was developed by Xerox PARC in 1970s, commercialized as a technology in 1980 and standardized by Institute of Electrical and Electronics Engineers (IEEE) in 1983 as IEEE 802.3. Ethernet has since undergone a lot of change to accommodate new features – such as, larger bit rate, higher number of nodes, improved range of communication, etc., and is now widely used in all forms of automation having wired communication. Ethernet technology first started using coaxial cable as the shared physical medium and later shifted to twisted pair copper cable. Many variants of ethernet like ethernet over power line and fiber optics exist for long distance and large bandwidth communication (Table 3.1).

TABLE 3.1
Types of Ethernet

Ethernet Type	Bandwidth	Cable Type	Maximum Distance
10Base-5	10 Mbit/s	Coaxial	500 m
10Base-2	10 Mbit/s	Coaxial	185 m
100Base-TX	10 Mbit/s	Twisted pair	100 m
100Base-TX	100 Mbit/s	Twisted pair	100 m
100Base-FX	200 Mbit/s	Twisted pair	100 m
100Base-FX	100 Mbit/s	Multi-mode fiber	400 m
1000Base-T	200 Mbit/s	Multi-mode fiber	2 km
1000Base-TX	1 Gbit/s	Twisted pair	100 m
1000Base-SX	1 Gbit/s	Twisted pair	100 m
1000Base-LX	1 Gbit/s	Multi-mode fiber	550 m
10GBase-CX4	1 Gbit/s	Single-mode fiber	2 km
10GBase-T	10 Gbps	Twin-axial	100 m
10GBase-LX4	10 Gbit/s	Twisted pair	100 m
10GBase-LX4	10 Gbit/s	Multi-mode fiber	300 m
10GBase-LR	10 Gbit/s	Single-mode fiber	10 km

Devices using ethernet for data communication divide larger data streams into smaller units called frames. Each of such frames will have a set of header control bytes like *source* and *destination* addresses followed by data *payload* and *error checking* data as tail end bytes of the frame. The end-to-end operation on ethernet for data communication is governed by the open systems interconnection (ISO-OSI) framework model of the International Organization for Standardization and later by the transmission control protocol and internet protocol (TCP/IP) model.

Coaxial cable: Coaxial cable has a central copper conductor that carries the data in the form of high frequency signals and usually has three more layers of insulation and shielding around it. The first layer is a dielectric insulator, typically PVC, and around this is the second layer of a metallic wrapping which acts as the second conductor for completing the circuit. This whole arrangement is covered in an insulating sheath or plastic cover. There are various standards that govern the choice and development of coaxial cables for real world applications (Figure 3.5).

Twisted pair copper cable: Twisted pair copper cable has two insulated copper wires of the same circuit twisted together which acts as conductors for data transmission. This twisting of wires of the same circuit will reduce electrical interference, noise from adjacent pairs and crosstalk besides improving electromagnetic compatibility. There are two variants of twisted pair copper cables – shielded and unshielded. Each conductor in the unshielded type has an insulation and then the complete set is wrapped by a sheath. A metal foil shielding is provided in the other type for individual or small group of twisted pairs and a collection of such pairs are wrapped around with braided mesh or metal foil or both (Figure 3.6).

3.2.1.1.1   ISO/OSI Model

OSI is a conceptual model that abstracts the nuances of specifics and technology of the telecommunication system and at the same time defines and standardizes its communication functions. The model divides a communication scheme into hierarchical abstraction layers to achieve the functionalities necessary for data

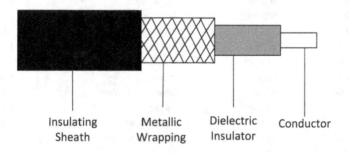

| Insulating | Metallic | Dielectric | Conductor |
| Sheath | Wrapping | Insulator | |

FIGURE 3.5   Coaxial cable.

FIGURE 3.6 Twisted copper pair cable.

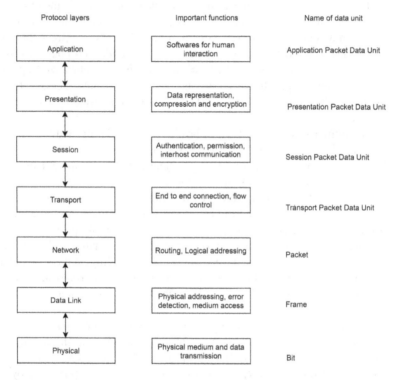

Protocol layers	Important functions	Name of data unit
Application	Softwares for human interaction	Application Packet Data Unit
Presentation	Data representation, compression and encryption	Presentation Packet Data Unit
Session	Authentication, permission, interhost communication	Session Packet Data Unit
Transport	End to end connection, flow control	Transport Packet Data Unit
Network	Routing, Logical addressing	Packet
Data Link	Physical addressing, error detection, medium access	Frame
Physical	Physical medium and data transmission	Bit

FIGURE 3.7 ISO/OSI protocol stack – layered architecture.

communication. It aims to facilitate interoperability of different communication schemes with standard communication protocols. The model has seven layers as shown in Figure 3.7.

Physical layer: The physical layer, hierarchically the lowermost layer of the model, realizes actual physical transmission and reception of raw information in the form of signals between a network device and the transmission medium. The layer functionally implements the system or devices that convert the digital bits representing data into electrical, radio or optical signals depending on the choice of the medium of communication. Various specifications and functionalities of the layer specify characteristics such as data rates, signal modulation strategy, power and voltage levels of the signal, timing of voltage changes, transmission range, channel access procedure and physical devices like connectors.

Data link layer: The data link layer provides functionalities for creating, maintaining and destroying links between neighboring nodes for node to node data transfer. The layer realizes features for framing, medium access, error control and flow control in addition to link management. Framing is the process of dividing the complete packet of data transferred from the network layer to smaller fragments so as to fit the buffer space in the physical layer. Medium access control is the process by which decisions as to how and when a node should access a medium for transmission or reception of data is decided. Error control includes error detection and possible error correction or retransmission. Flow control involves synchronized operation, in terms of data transfer rate and frequency of data message flow, among neighboring nodes. IEEE 802 divides the data link layer into two sublayers:

1. Medium access control (MAC) layer – responsible for medium access control and assigns MAC or physical address.
2. Logical link control (LLC) layer – responsible for framing, error control, flow control and link management.

Network layer: The network layer provides functionalities for sending data available in this layer in the form of packets from one node to another node in the same or different network. The process of sending packets from a source node to a destination node by providing the data and the addresses is called *routing*. A packet can have varying size depending on the amount of data pushed down from the transport layer. Another important functionality of the network layer is logical or IP address allocation, management and deallocation.

Transport layer: It provides functionalities for transferring data sequences of varying length from a source node to a destination ensuring quality of service. This is achieved by having protocols for end-to-end delivery of packets, flow control and error control. The two main schemes employed in transport layer are transmission control protocol (TCP), which is connection oriented and hence acknowledgement based, and user datagram protocol (UDP), which is connectionless and uses no acknowledgement. The transport layer does segmentation and reassembly of messages in which long messages from the application layer are divided into smaller messages; whereas smaller packets from the network layer are combined to form larger messages. The data is represented in the form of segments or datagram.

Session layer: The session layer provides functionalities for establishing, managing and terminating the connections between applications on the local as well as remote devices. It facilitates procedures for starting check pointing, suspending, restarting and terminating a session in full-duplex, half-duplex, or simplex operation modes.

Presentation layer: The presentation layer provides functionalities to set up a link between application layer services of the source and destination devices; it also establishes a mapping for the application layer services to use varying syntax and semantics. This layer represents data as presentation protocol data units which are packed into session protocol data units and sent down the OSI protocol stack. The presentation layer acts as data translator between the application and the network layers' data formats.

Application layer: The application layer, the topmost layer of OSI stack, is the direct user interface with the communication system. This layer provides functionalities that enable interaction of user side software applications and systems and software that implement a communicating component. This is achieved by identifying communication peers, evaluating resource availability, and providing communication synchronization.

3.2.1.1.2 TCP/IP Model

The TCP/IP stack has only four hierarchical layers in achieving all the communication functionalities necessary for networking multiple devices for fruitful exchange of data. The functionalities of the physical and data link layers of OSI are provided in TCP/IP in the host to network interface layer. The functionalities of network and transport layers of OSI are provided in TCP/IP in the internet and transport layers respectively. Functionalities of the top three layers of OSI – namely, the application, presentation and session – are achieved using the application layer in TCP/IP model (Figure 3.8).

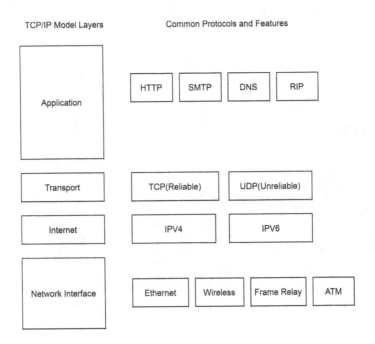

FIGURE 3.8   TCP/IP protocol stack – layered architecture.

3.2.1.1.3 Network Topologies

The main purpose of any network technology is to realize multipoint to point, point to multipoint or multipoint to multipoint connectivity and message passing in addition to the straightforward point to point communication. Ethernet in this effort allows the following topologies for interconnection of devices for realization of a network (Figure 3.9).

Bus topology: A single central cable acts as the shared medium in the bus network and interconnects all members of the network. Any member node of the network can transmit a data packet to any other node by sending the data to the bus. Though all connected nodes can receive the data, only the intended receiver will process it.

Ring topology: In the ring network topology, every member node is connected to two other nodes such that all of them together form a ring, as the last node will be connected to the first one. The messages can travel in clockwise or anticlockwise direction.

Star topology: Every member node is connected to a common node forming a star with one central point and several end points, in star network topology. The communication happens between the central node and the end nodes; there is no provision for direct communication between the end nodes.

Mesh topology: All member nodes in mesh network (commonly known as the full mesh) topology are connected to each other in highly redundant fashion. When some member nodes are connected to multiple nodes, but not to all other nodes of the network, it is called as partial mesh topology. This topology provides multiple paths for data flow between one node and another.

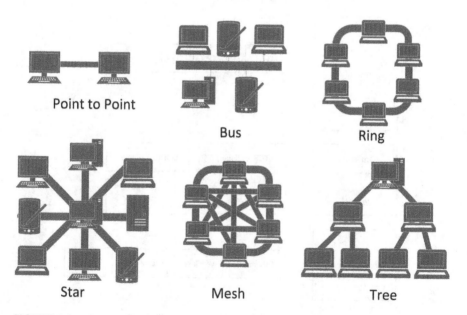

FIGURE 3.9 Network topologies in Ethernet.

Tree topology: A tree or hierarchical topology is a specialized network topology in which every parent node has two or more child nodes connected to it and a single root node forms one end point of the network. It is a combination of bus and star topologies. This is a hybrid topology and there exist large possibilities for creating various hybrid topologies by combining the above-mentioned topologies or improvising on one.

3.2.1.1.4 Network Devices and Terminologies

Any communication network consists of a large number of nodes with multiple functions to fulfill the requirements for which the network is constituted. The following are some of such specialized functional nodes in ethernet network (Figure 3.10).

Repeater: A repeater is a device that is used to boost the power level of the signal by regenerating it when the signal power level becomes too weak after long distance transmission. It is a two port device that operates in the physical layer and helps to increase the range of the network.

Hub: A hub is basically a multiport version of a repeater. The difference between hub and repeater is the multiple input/output lines available in a hub. Any signal that is received in any line of a hub is regenerated and sent to devices connected to all other lines of the same hub. Both hub and repeater are not intelligent devices and do not read the data frames, instead they just repeat sending these to all lines except the incoming one.

Bridge: A bridge can be considered as two port intelligent repeater as it can read the MAC address and pass the incoming message to the outgoing line. As it works looking at the MAC address it is operating in the data link layer. A bridge can interconnect two similar networks.

Switch: A switch is a multiport bridge with a buffer that can perform error checking before forwarding the data according to the MAC address. It can interconnect various similar networks and traditionally works in data link layer.

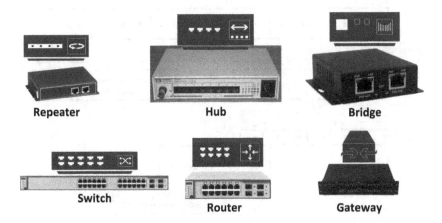

Repeater   **Hub**   **Bridge**

Switch   **Router**   **Gateway**

FIGURE 3.10 Network devices in Ethernet.

There are switches that incorporate some routing strategies and hence work in the network layer.

Routers: A router is a device that operates in the network layer which routes data packets based on the IP address. Routers interconnect multiple similar networks together and dynamically update routing table based on which decisions are made as which packet is to be forwarded to which node or network.

Gateways: Gateway is a network device used to interconnect two or more networks operating on different protocols. These are basically protocol converters and can operate in any layer of the protocol stack and act as a bridge between multiple networks by interpreting the address on the packet and forwarding to the correct destination node or network.

3.2.1.1.5   MAC Address

MAC address is a 48 bit unique identifier assigned to all network interface controllers of all network devices. MAC address is usually assigned by device manufacturers and hence is referred to as burned-in address or physical address. The address includes an organizationally unique identifier represented by the first three octets and a network interface controller address represented by the last three octets (Figure 3.11).

Addresses can be administered universally or locally. In the universal case, the manufacturer assigns the unique address. The first three octets are called the *organizationally unique identifier* and are used to identify the manufacturer. The remaining three octets can be assigned by the organization in any manner they choose, but the entire number must maintain its uniqueness. A locally administered address is usually assigned by the network administrator or user by overriding the original physical address. If the second least-significant bit of first octet is zero, the address is

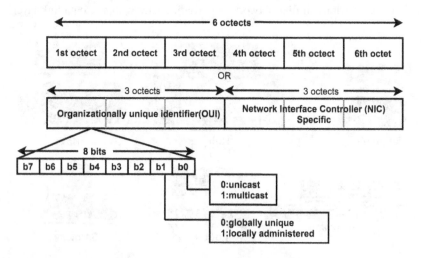

FIGURE 3.11   MAC address.

universally administered and if this bit is one, the address is locally administered. If the least-significant bit of the first octet is set to zero, then the frame is supposed to reach only one receiving network interface card (NIC) realizing unicast addressing and if this bit is set to one, then it is multicast addressing.

3.2.1.1.6 IP Address

Internet Protocol (IP): Internet protocol address or the logical address of any device on a network following internet protocol shows host or network identification and location. Unlike MAC address, IP address is not unique globally. IP address specifically identifies the network in which a node is present and also the location of the specific node in the network. IP address can be assigned to a device in a network either by the network administrator which can be static or dynamic or by the user which is static. It is used in the network for routing the packets so as to guide it from source node to destination node. The router or gateway has a routing table which has IP addresses to help routing. Two versions of IP addressing are in use today.

IP version 4: IP version 4 uses 32 bit addressing which allows 2^{32} devices to be part of the network with unique address. In this, some address blocks are reserved for private networks and multicast addresses. The IP address consists of two parts: the network identifier and the host identifier. There are two defined styles for IPv4 addressing – namely, *classful* and *classless* inter domain routing. In classful addressing, network classes are created for address definition and one, two or three octets are allocated for network or subnet addressing and the remaining for host addressing. As this scheme limits the possible maximum network addresses to 2^{21}, a new addressing scheme was developed that allocates address to networks and end users or hosts on any address bit boundary. The 32 bits of IPv4 is usually represented in four octets separated by dots and each octet value is written in decimal form (Figure 3.12).

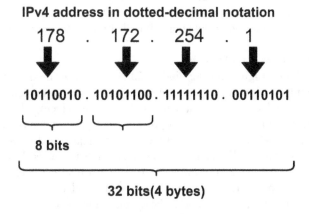

FIGURE 3.12 IPv4 address representation.

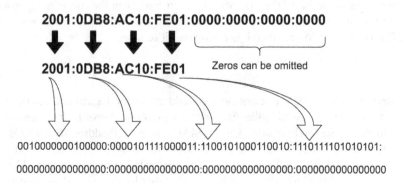

An IPV6 address (in hexadecimal)

2001:0DB8:AC10:FE01:0000:0000:0000:0000

2001:0DB8:AC10:FE01   Zeros can be omitted

0010000000100000:0000101111000011:1100101000110010:1110111101010101:

0000000000000000:0000000000000000:0000000000000000:0000000000000000

FIGURE 3.13   IPv6 address representation.

IP version 6: This style of addressing uses 128 bits thereby allowing 2^{128} devices with unique addresses to be part of the network. Thus, its address space is large compared to that of IPv4. IPv6 is represented as eight groups with values separated by full colons while each group has 16 bits written in hexadecimal form in lower case. IPv4 addresses are mapped to IPv6 by making the least significant 32 bits of IPv6 the same as IPv4 address and the remaining most significant bits in IPv6 format. IPv6 uses the last 64 bits always as interface address (Figure 3.13).

3.2.1.2   Fiber Optic Communication

In fiber optic communication, information is exchanged between two devices by transmitting pulses of light, typically infrared, through optical fiber cable. Light is the electromagnetic carrier wave for data transmission and is modulated to carry original data. Fiber optic cable has high bandwidth and better immunity to electromagnetic interference; it provides longer distance compared to coaxial or twisted copper pair cable. Researchers at Bell Labs (Alcatel Lucent) have achieved speeds of over 100 Petabits per second kilometer using fiber optic communication.

1. *Components and Techniques*
   Transmitters: Light emitting diodes (LEDs) and laser diodes are the most commonly used optical transmitters in fiber optic communication. Power, speed, ruggedness, and cost are among the common factors that help frame the selection of transmitter. The optical transmitters used must be compact, efficient and reliable, while operating under standard conditions. The LEDs produce incoherent light and are used in small distance and low bandwidth applications. Surface emitting types of LEDs are simple and reliable, but they have limited frequency modulation and broader spectral width. Edge emitting LEDs have overcome these limitations and have better power rating. Laser diodes, thanks to high power, speed and narrower spectral

bandwidth characteristics are preferred for long distance and high data rate applications; the disadvantage is that they are nonlinear and sensitive to temperature variations.

Receivers: A photodiode, typically a semiconductor-based photodetector, is the main part of an optical receiver which converts light into electricity with the photoelectric effect. Various types of photodiodes include p-n photodiodes, p-i-n photodiodes, and avalanche photodiodes and metal-semiconductor-metal photodetectors.

Low Loss Optical Fiber: Fiber optic cable is made of high-quality silica glass or plastic; it is flexible and contains one or more optical fibers. The three main types of optical fiber cable are, single-mode fiber (SMF), multi-mode fiber (MMF) and plastic/polymer optical fiber (POF). The SMF allows at a time one mode of light to propagate thorough its small core and requires costlier components and interconnection systems, but on the brighter side allows much lengthier and better performing links. MMF having a larger core allows propagation of multiple modes of light at a given instant and needs less precise and cheaper transmitters, receivers and connectors. The disadvantage of MMF is that it introduces multi-mode distortion, resulting in limited bandwidth and distance of the link. Furthermore, they are usually expensive and exhibit higher attenuation. POF replaces the glass in conventional optical fiber with plastic or polymer. POF has higher degree of robustness under bend, stretch or stress, much larger core diameter and lower cost. The disadvantage is that it supports lower data rates and distance.

A fiber optic cable has its core made of glass or plastic, cladding facilitating total internal reflection, plastic coating, strengthening fibers and outer jacket. The core is a cylindrical shaped strip whose diameter depends on the application. As a result of the internal reflection, the light moving inside the core reflects at the core-cladding boundary. Cladding is again an optical material that surrounds the core and reflects light back to the core. When light from the dense core enters into the less dense cladding, it gets reflected back to the core. The plastic coating surrounding the cladding acts as a reinforcement to the core in case of shocks and excessive bends in cable. The strengthening fibers are another layer of protection against excessive forces on the fiber, particularly during installation. The outer jacket is the sheath to protect the cable from environmental factors. The jacket is usually color coded to distinguish between various types and applications (Figure 3.14).

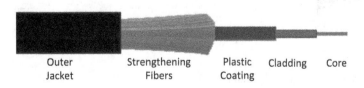

| Outer | Strengthening | Plastic | Cladding | Core |
| Jacket | Fibers | Coating | | |

FIGURE 3.14 Fiber optic cable.

Wavelength division multiplexing (WDM) is a technique in which multiple light beams of different wavelength, each modulated with separate information, are transmitted through a single optical fiber to form multiple channels of information. This in effect multiplies the available capacity of optical fibers. To implement WDM, a wavelength division multiplexer at the transmitter side and a de-multiplexer at the receiver side are required; both are satisfied by *arrayed waveguide gratings*.

A fiber optic communication system has interfacing circuits at transmitting and receiving points, light source and light detector as shown in Figure 3.15. With the input data given, the interfacing circuit at the transmitter side together with the light source generates light beam according to the received electrical signals. The fiber optic cable carries this light beam to the destination where the receiver-side photo detector along with an appropriate electronic circuit converts the information back to the original data bits (Figure 3.15).

The distance over which a fiber optic communication system can transmit data is influenced by fiber attenuation and fiber distortion among other factors. Solutions to these problems are *optoelectronic repeaters* and *optical amplifiers*. The optoelectronic repeaters convert the light signal into an electrical signal, boost it and then send it using an optical transmitter at a higher intensity than it was received. Optical amplifiers amplify the optical signal directly without converting it. A common optical amplifier is an erbium-doped fiber amplifier (EDFA) developed by doping a fiber with erbium. It sends light from a laser with a smaller wavelength than the communication signal.

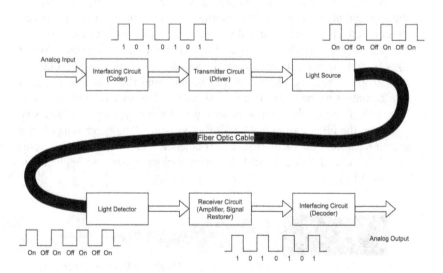

FIGURE 3.15   Fiber optic communication.

2. *Standards and Specification*: There exist several industrial standards – such as, enterprise system connection (ESCON)/serial byte connection (SBCON), fiber distributed data interface (FDDI), synchronous optical network (SONET) and Gigabit Ethernet – for realizing an optical fiber communication network. The ESCON/SBCON, first introduced in 1990 by IBM and adopted as SBCON by American National Standards Institute (ANSI) in 1996, is a point to point, bidirectional, fiber optic data link having maximum data rate of 17 Mbyte/s. The ESCON devices can communicate either directly via a channel-to-channel connection, or via a central non-blocking dynamic switch. An ESCON data frame includes a two-byte start of frame delimiter, destination and source address of two-bytes and one byte of link control information for connection request to form the header, payload of data up to 1028 bytes, a trailer of two-bytes cyclic redundancy check (CRC) for errors and an end of frame delimiter of three-bytes. FDDI, proposed in 1980s by ANSI, is a standard for data transmission in LAN – one of the pioneers to specify optical fiber as the physical medium. The FDDI confirms to the architectural concepts of OSI and uses a logical topology of ring-based token network, developed from timed token protocol in IEEE 802.4 token bus. The FDDI names four layers – the physical layer, physical media dependent, media access control and station management. FDDI has bandwidth of 100 Mbit/s, data frame of 4500 bytes, range up to 200 km and has two rings; the second one is a backup if the primary ring fails. FDDI was replaced by Fast Ethernet and later by Gigabit Ethernet. SONET defined by Telcordia and ANSI and synchronous digital hierarchy (SDH) defined by European Telecommunications Standards Institute (ETSI) are the standards used for synchronous transmission of multiple digital bit streams through optical fiber. The framing followed in SDH is synchronous transport module level 1 (STM-1) which has a bandwidth of 155.520 Mbit/s with a frame size of 2430 octets. Nine octets of header termed as overhead is transmitted followed by 261 octets of payload or data and this cycle is repeated nine times to complete 2430 octets in 125 μs. The SONET offers an additional basic framing called synchronous transport signal, base level (STS-1) which has a bandwidth of 51.84 Mbit/s with a frame size of 810 octets. Three octets of header termed as overhead is transmitted followed by 87 octets of payload or data and this cycle is repeated nine times to complete 810 octets in 125 μs. Every SDH/SONET connection uses at least two optical fibers. Fast Ethernet increased the data rate from 10 to 100 Mbit/s and Gigabit Ethernet is introduced by IEEE in 1998 as IEEE 802.3z and further ratified as IEEE 802.3ab in 1999 increasing the data rate to 1000 Mbit/s. There are many physical layer standards defined for use in Gigabit Ethernet which uses optical fiber, twisted pair cable and shielded balanced copper cable. IEEE 802.3z standard includes 1000BASE-SX for transmission over MMF and 1000BASE-LX for transmission over SMF along with other definitions (Table 3.2).

TABLE 3.2
IEEE Fiber Optic Standards

Standard	Data Rate (Mbit/s)	Cable Type	Max. Distance
10Base-FL	10	Multi-mode	2 km
100Base-FX	100	Multi-mode	2 km
100Base-SX	100	Multi-mode	300 m
100Base-LX	100	Single-mode	100 km
1000Base-SX	1000	Multi-mode	220–550 m
1000Base-LX	1000	Multi-mode & Single-mode	550 m–2 km
1000Base-LH	1000	Single-mode	70 km

3.2.1.3 Power Line Carrier Communication – IEEE 1901–2010

Power line carrier communication (PLCC) uses the conductors of the regular electrical power transmission and distribution network as the physical medium for data transmission. A variety of power line communication technologies with different data rates and frequencies are needed for different applications, ranging from smart metering to transmission and distribution grid automation.

A typical PLCC network scheme used in power substations is shown in Figure 3.16.

1. *Components and Techniques*: The system has mainly three parts – namely, line or wave trap, coupling capacitor, and line tuner. *Line trap* is a parallel L-C filter (normally a band stop filter) that blocks the carrier signal frequencies and allows the power system frequency to pass through and is connected in series with the transmission line. *Coupling capacitor* acts as the physical interconnection between the electric transmission or distribution line and the PLCC transceiver for relaying the carrier signals. It provides high opposition to power frequency and low opposition to carrier signal frequencies. *Line tuner* is connected with the coupling capacitor to form a high pass or band pass filter to pass the carrier signal frequency. It provides

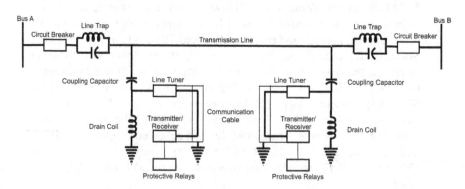

FIGURE 3.16 Power line carrier communication.

impedance matching for the PLCC terminal with the transmission or distribution line so as to push the carrier frequency to the electric power line. *Drain coil* is provided to filter out any power system frequency still present (Figure 3.16).

The generic PLCC system uses various frequencies like low and medium frequency (24–500 kHz), high-frequency ($\geq$1 MHz), ultra-high frequency ($\geq$ 100 MHz) for transmission of data over power line for a variety of applications (Table 3.3). The data rates supported vary from a few hundred bits per second to several megabits per second and under special circumstances even up to gigabits per second. There is a wide range of applications that use PLCC outside the scope of smart grid like telemetry, telephony and home automation and networking. Smart metering, home appliance control, protective relay, inter substation and control station communication and wide area monitoring and control are some applications that use PLCC within smart grid domain. The PLCC standards for home applications include X10 protocol and HomePlug; HomePlug and IEEE 1901 are used for long distance communication which includes features from HomePlug and HD-PLC.

2. *Standards and Specification*: The IEEE 1901-2010 uses frequencies below 100 MHz and has a data rate up to 500 Mbit/s and is used by various broadband over power line (BPL) devices including those used within buildings for LAN, smart energy applications and other smart grid data communication applications. IEEE 1901 supports two physical layers, one being FFT orthogonal frequency division multiplexing (OFDM) modulation derived from HomePlug and wavelet OFDM modulation derived from HD-PLC. Above these two physical layers, two MAC layers are defined to meet the requirements of (i) in-home network and (ii) internet access. IEEE 1901 supports time division multiple access (TDMA) and carrier sense multiple access/collision avoidance (CSMA/CA) to do medium

TABLE 3.3
PLC Technology Specification

	Low Data Rate	Medium Data Rate	High Data Rate
Standards	IEC 61334, ANSI/EIA, 709.1, .2, UPB	PRIME, G3, P1901.2	G.hn, IEEE 1901
Modulation	BPSK, FSK, SFSK, QAM	PSK+OFDM	PSK+OFDM
Data rate	0–10 kbit/s	10 kbit/s–1 Mbit/s	Greater than 1 Mbit/s
Frequency range	Up to 500 kHz	Up to 500 kHz	MHz
Application	Control and command	Control, command and voice	Broadband over power line

access control. Inter-system protocol (ISP) manages the various devices and systems connected via PLCC for the coexistence of different physical and MAC layers and also prevents interference of various devices operated in close proximity.

The X10 protocol developed in 1975 by Pico Electronics uses the existent power lines as the physical medium for communication between smart electronic devices. The data to be communicated is encoded onto a 120 kHz carrier signal that is transmitted as bursts during the zero crossings of the AC power signal with a single bit at each zero crossing. The X10 data exchange happens with four bits house code followed by four bits unit code and finally four bits command code. Usually, X10 devices operate like a controller of operations of a device and there is only one-way communication. Provision exists for two-way message exchange in X10 where the controller queries about the status of a device and the device responds. In X10, bit "1" is transmitted with an active zero crossing (a one millisecond burst of 120 kHz at zero crossing) followed by an inactive zero crossing (no pulse at zero crossing). Bit "0" is transmitted by an inactive zero crossing followed by an active zero crossing. The data frame format includes a start code having three active zero crossings and an inactive zero crossing, four bits house code and five bits function code. The first four bits of the function code form the unit code or command and 0 in the last bit indicates a unit code, whereas 1 indicates a command. Multiple unit codes can be sent before finally sending a command to control multiple devices in the same way. All frames are sent twice, and the effective data rate is around 20 bits per second.

HomePlug is the family name that refers to different PLCC specifications such as low data rate internet content, low power, low throughput, applications such as smart power meters and in-home data exchange between electric systems and appliances. All commercial HomePlug devices satisfy the advanced encryption standard (AES)-128 encryption standard. Most HomePlug devices act as adapters, which are plugged into wall power outlets and provide one or more Ethernet ports. End devices with functionalities found in such standalone adapters are also evolving. This allows a smart device to use wired Ethernet, powerline or wireless communication to facilitate a redundant and reliable failover system in realizing smart homes.

HomePlug 1.0, introduced in 2001, provides a peak physical layer data rate of 14 Mbit/s, has since been replaced by HomePlug AV which offers a peak data rate of 200 Mbit/s at the physical layer and about 80 Mbit/s at the MAC layer. The HomePlug AV2 specification, brought out in 2012, is interoperable with all previous versions and is IEEE 1901 compliant. HomePlug Green PHY with peak rates of 10 Mbit/s falls under HomePlug AV and is intended for use in smart grid applications such as smart meters and small end appliances so as to facilitate data exchange over home network and with the power utility. HomePlug devices normally do not coexist with devices that employ other powerline technologies, but IEEE 1901 with its ISP enables coexistence of all PLCC systems.

3.2.2 Wireless Communication Technologies

The term wireless communication covers all types of message passing between any two (or among multiple) devices using a radio signal with air as the physical medium of communication. A broader classification based on means of wireless communication achieved can include radio-based, free space optical, sonar and electromagnetic induction, out of which only radio-based technologies are discussed here because of the obvious limitations in the other systems for meeting of smart grid communication requirements. If the communication is exclusively between two devices, it is *point to point* communication. If the communication involves multiple devices, then it can be *point to multipoint, multipoint to point* or *multipoint to multipoint*, which is a mesh network. There exist a number of technologies that use electromagnetic waves to transmit data over air and a lot of features, like the frequency spectrum, differentiate all these technologies. The various available wireless technologies differ in many factors including availability, range and performance. If a wireless communication network employs only one technology, then it is called a *homogeneous* network. In some cases, to meet the application requirements, devices supporting multiple wireless technologies are interconnected to form *heterogeneous* networks. The main advantages of wireless communication include less installation and maintenance cost, easy access, setup and removal, scalability and mobility. Another outstanding feature is the wireless multicast advantage, which is the capability of a wireless node to reach out to many of its neighboring nodes with single transmission – this is extensively exploited in resource constrained systems like wireless sensor networks to save energy. The main disadvantages of wireless communication include security threats, range and issues on quality of service.

3.2.2.1 Cellular Communication

A cellular or mobile network communication is a system in which the last link involves wireless data exchange between a fixed device and various mobile devices.

1. *Components and Techniques*: A cellular network has the following structure.
   a. Mobile devices
   b. Base station network
   c. Circuit switched network and Packet switched network
   d. Public switched telephone network

   This network is widely distributed, and the distributed mobile devices are grouped under cells which are served by at least one fixed location device called the base station. Base station provides network coverage to these cells and these base stations are connected together to form base station network. Each cell uses a frequency that is different from the neighboring cells. All such cells together can serve a very large geographic area. This enables practically any mobile device in any cell to talk to any other mobile device in the same or different cell. This communication, irrespective of the cell to which the mobile devices belong to, happens via the base station and sometimes via a master switching center. The mobile switching center has *circuit*

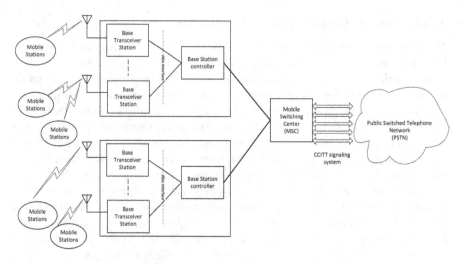

FIGURE 3.17 Cellular network architecture.

switched network for handling voice calls and text, as to be communicated as part of the cellular communication typically handled under global system for mobile (GSM) communications and *packet switched network* for handling data packets typically handled under general packet radio service (GPRS). The public switched telephone network is used to connect devices to the wider cellular network (Figure 3.17).

2. *Standards and Specification*: The GSM is a trademark of GSM association and was developed by the European Telecommunications Standards Institute (ETSI) in 1980s and first deployed in 1991 to replace analog cellular networks. GSM presents a digital circuit switched network for full duplex voice exchange and then expanded for data exchange with packet data transport via GPRS and enhanced data rates for GSM evolution (EDGE). 2G GSM uses 900 MHz and 1800 MHz normally and 850 MHz and 1900 MHz are used in some countries and very rarely 400 MHz and 450 MHz are also used. GSM uses TDMA, frequency division multiplexing and its variants for resource allocation among various connected devices. Code division multiple access (CDMA) is another 2G technology that uses code division system for resource allocation. The CDMA is a more powerful and flexible technology, but it does not support voice and data transfer at the same time. The GPRS is packet data transfer standard for 2G and 3G networks of GSM developed by ETSI and managed by 3rd Generation Partnership Project (3GPP). GPRS achieved data rates of 56–114 kbit/s in 2G. GPRS performance varies based on the number of connected users and it is a best effort service. GPRS supports some additional protocols, for example, IP, point to point protocol (PPP), and X.25 – and provides an array of extra functions in the GSM network. EDGE developed by AT&T in 2003 and standardized by 3GPP through better methods of coding and data transmission allows improved bit

rates per channel (up to 1 Mbit/s and typically 400 kbit/s) on GSM network and can be seen as development on the GPRS system. The universal subscriber identity module (USIM) and several stream ciphers like A5/1, A5/2 and A5/3 and GEA/1-4 are some measures available for providing security to data and operation in GSM and GPRS networks. The identity of a user is verified in GSM network usually through a challenge response mechanism using a 128 bit challenge and a 32 bit signed response.

Third generation (3G) of wireless mobile communication is an improvement over the 2G system; it provides better connectivity and voice and data transfer functions with an improved data transfer rate from a minimum of a few hundred kbit/s to a few Mbit/s which varies according to the underlying technology implemented over 3G. Universal mobile telecommunications system (UMTS) is a 3G technology that offers better spectral efficiency and bandwidth by defining a complete set of radio network, core network and authentication of users demanding new base stations and frequency bands. Wideband code division multiple access (WCDMA), an air interface standard in 3G, uses the direct sequence spread spectrum (DSSS)-based CDMA channel access method with a pair of 5 MHz wide channels. High speed downlink packet access (HSDPA) and high speed uplink packet access (HSUPA) combine to form high speed packet access (HSPA) that improves the performance of WCDMA-based 3G systems and provides up to 14 Mbit/s in the downlink and 5.76 Mbit/s in the uplink. Evolved high speed packet access (HSPA+), which is an improvement over HSPA released in 2008, allows data rates as high as 168 Mbit/s in the downlink and 21 Mbit/s in the uplink.

Long term evolution (LTE) is a standard for mobile wireless communication technology, developed by 3GPP, with improvements in radio interface and core network technologies and is based on GSM/EDGE and UMTS/HSPA. The LTE specification, proposed in early 2000s and finalized in late 2000s, offers peak data rates of 300 Mbit/s for downlink, 75 Mbit/s for uplink and a peak transfer latency of 5 ms. The LTE supports frequency division duplexing and time division duplexing and uses an IP-based network architecture to replace the GPRS core network. LTE is an upgradation of 3G UMTS to pave way for 4G by improving and modifying the combined circuit and packet switched network to an all IP system. LTE-Advanced (LTE-A) or True 4G, proposed by NTT DoCoMo is the fourth-generation cellular communication technology that can reach the original peak recommended speeds of 100 Mbit/s for high mobility communication and 1 Gbit/s for low mobility communication. Unlike previous generation technologies 4G does not support traditional circuit switched network, but on the other hand uses an all IP-based communication system. The fifth-generation wireless technology for digital cellular communications, 5G, divides the frequency spectrum into *millimeter waves, mid-band* and *low-band*. Millimeter wave is the fastest with speeds in the range of 1–2 Gbit/s and mid-band (the most widely deployed) with speeds in the range 100–400 Mbit/s and low-band offering similar capacity as 4G with an air latency of 8–12 ms. The 5G is expected to grow even up to 100 Gbit/s (Table 3.4).

TABLE 3.4

Cellular Technology Generations – Specifications

Technology	1G	2G	3G	4G
Interface	FDMA	TDMA/CDMA	WCDMA, CDMA, etc.	OFMD, OFMDA, MIMO, etc.
Throughput	<400 kbit/s	600 kbit/s	Above 2 Mbit/s	Up to 200 Mbit/s
Coverage	<10 km	<10 km	Up to 10 km	Above 50 km
Frequency	150 MHz	300–600 MHz	900, 1800, and 2100 MHz	2–8 GHz

3.2.2.2   Wi-Fi – IEEE 802.11-2016

Wi-Fi, a trademark of the Wi-Fi Alliance and standardized based on IEEE 802.11 family of standards, is a wireless local area network that provides connectivity for data exchange and internet access with comparable link speeds as wired LAN. Wi-Fi compatible devices can connect to an access point to communicate to another device as well as to wired devices and even to be part of the Internet. Wi-Fi uses the 2.4 and 5 GHz open band known as the Industrial Scientific and Medicinal (ISM) band which are further subdivided into different channels. An access point typically has a range of 20 m indoors while some access points claim to have a range of 150 m outdoors. The first version of 802.11 protocol came out in 1997 with a data rate up to 2 Mbit/s; currently the maximum data rate supported by Wi-Fi extends to a few Gbit/s.

1. *Components and Techniques*: Devices of a Wi-Fi network can operate in two different modes – namely, *infrastructure mode* and *ad-hoc mode*. The most commonly used one is infrastructure mode in which all nodes communicate via an *access point* like a base station in GSM network. Any node can talk to any other node in the infrastructure mode, provided these are connected to some base station and are parts of some interconnected network. Wi-Fi devices can communicate directly to each other without passing data through the access point in the ad-hoc mode and in its complex versions, even far off source and destination can and do pass messages via intermediate nodes. Typically, ad-hoc mode is employed to create temporary communication networks using Wi-Fi. An access point is a Wi-Fi network device that provides interconnection capability to member nodes to a common Wi-Fi and another wired network and is different from a Wi-Fi hotspot which is a physical location with Wi-Fi access to network (Figure 3.18).

   A device normally needs to know *network name* and *password* to get connected to a Wi-Fi network. The network name is called a service set identifier (SSID) which is 32 byte long; it identifies a network and also a service set. A service set is the set of the devices associated with a specific Wi-Fi network. A *basic service set* (BSS) is a group consisting of

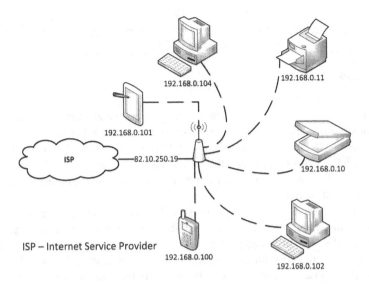

FIGURE 3.18   Wi-Fi network.

devices and base station where all members share the same physical layer and medium access properties. Here, each BSS is recognized by a MAC address which is called the BSSID. If the BSS is formed in ad-hoc mode, then it is called *Independent BSS*. A collection of BSS that share the same SSID is known as the extended service set (ESS). A lot of versions and standards exist for Wi-Fi devices and since inception Wi-Fi devices are listing the IEEE standards supported by these. Wi-Fi alliance started generation numbering from 2018.

2. *Standards and Specification*: A list of important Wi-Fi generations, associated IEEE standards and specifications are given in Table 3.5.

   Wi-Fi devices communicate like any other radio by sending data packets as in wired LAN with 802.3. The various protocols implemented at different layers of the protocol stack varies with versions of Wi-Fi. 802.11

TABLE 3.5
IEEE Standards and Specifications for Wi-Fi

Generation/IEEE Standard	Maximum Data Rate	Adopted	Frequency
Wi-Fi 6 (802.11ax)	9608 Mbit/s	2019	2.4/5 GHz 1–6 GHz ISM
Wi-Fi 5 (802.11ac)	6933 Mbit/s	2014	5 GHz
Wi-Fi 4 (802.11n)	600 Mbit/s	2009	2.4/5 GHz
802.11g	54 Mbit/s	2003	2.4 GHz
802.11a	54 Mbit/s	1999	5 GHz
802.11b	11 Mbit/s	1999	2.4 GHz

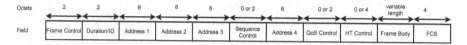

FIGURE 3.19   Wi-Fi frame format.

provides a wide variety of frequency range for use and each range is divided into multiple channels. Spectrum allocation, operational criteria and limitations vary from country to country. 802.11b/g/n uses the 2.4 GHz ISM band and 802.11a/h/j/n/ac/ax uses the 5 GHz band. The data in Wi-Fi is arranged as 802.11 frames much similar to Ethernet frames at the data link layer with extra address fields (Figure 3.19).

The channels are used in half duplex mode and are accessed in a timeshared fashion by multiple devices. When a device communicates to another, all the neighboring nodes within the range receive the data, even when the information is sent to only one destination. Devices are normally programmed with a globally unique 48 bit MAC address. Wi-Fi uses destination and source addresses to create link level connections and on receiving a packet, the receiver uses the destination address on the packet to determine whether the packet is for the station or not. As the channel is shared, when two stations attempt to transmit data at the same time, it most often results in collision of packets in the medium. This corrupt the transmitted data sometimes even beyond the possible extent of correction and needs re-transmission. The data loss and re-transmission of packets have adverse impact on performance metrics like the *packet delivery ratio* and *throughput*. Accurate transmission is not guaranteed because of significant interference and large probability of collisions in the wireless medium and Wi-Fi is treated as a best effort delivery system. Wi-Fi's MAC layer traditionally uses CSMA/ CA and its variants to reduce the collisions, access the medium and manage retries without relying on higher levels of the protocol stack.

Security is an important aspect of Wi-Fi networks, as is the case in any wireless communication system. In a wireless communication system any device in range can access the network, unlike the physical restrictions that are present in a wired system. So, several measures have been incorporated into various versions of Wi-Fi to enhance data and network security and prevent unauthorized access of data. Some common methods to provide basic security include mandating a password along with the SSID to join any Wi-Fi network, hiding the access point name by disabling SSID broadcast, and allowing only devices with known MAC addresses to join the network. But none of these are foolproof security measures. Wired equivalent privacy (WEP) was introduced in Wi-Fi networks to provide data confidentiality; it used RC4 stream cipher for confidentiality and the CRC-32 checksum for integrity. Standard 64 bit WEP uses a 40 bit key with a 24-bit initialization vector (IV) to form the RC4 key and a 128 bit WEP uses a 104 bit key with a 24-bit IV. Wi-Fi protected access (WPA), WPA2 and

WPA3 are security protocols developed for securing the information moving across and operation of Wi-Fi networks. WPA employs the temporal key integrity protocol (TKIP) which dynamically generates a new 128-bit key for each and every packet and has a message integrity check with TKIP which replaced the CRC. WPA2 brought in support for AES-based encryption. WPA3 uses a 192 bit encryption and also replaces the pre-shared key exchange with simultaneous authentication of equals providing a secure initial key exchange.

3.2.2.3   WiMAX – IEEE 802.16-2009

Worldwide interoperability for microwave access, popularly known as WiMAX, is a wireless broadband communication technology similar to Wi-Fi, but supports higher data rates and better range with a greater number of users. WiMAX is based on IEEE 802.16 standards released in 2005 and provides multiple physical and MAC layer options. Even though WiMAX is used for all devices following 802.16 standards, it applies only to interoperable devices that satisfy specific conformance criteria mandated by the WiMAX Forum.

1. *Components and Techniques*: A WiMAX system consists of two major parts:
   a. A WiMAX base station
   b. A WiMAX receiver
   A WiMAX base station consists of all electronic circuitry and a tower and uses the defined MAC layer as per the standard to create a common interface to make networks interoperable. It allocates uplink and downlink bandwidth according to the needs of subscribers. Each tower covers an area called a cell, theoretically up to 50 km and practically about 10 km. A WiMAX receiver could be a stand-alone box or a NIC connected to the device as a part of the WiMAX network. WiMAX provides two forms of wireless connectivity – namely, non-line of sight (NLOS) and line of sight (LOS). In both the WiMAX antenna on the device connects to the WiMAX tower (Figure 3.20). The device and tower are out of line of sight and the frequency range used is 2–11 GHz in NLOS. The device antenna in LOS points straight at the WiMAX tower and uses higher frequencies up to 66 GHz.

2. *Standards and Specification*: The physical layer uses scalable orthogonal frequency-division multiple access (SOFDMA) with 256 sub-carriers offering good resistance to multipath noise. More advanced versions of WiMAX bring in multiple antenna support through multiple-in multiple-out (MIMO) with benefits in terms of range, bandwidth, power consumption and frequency reuse (Table 3.6). The peak physical layer data rate can be as high as 74 Mbit/s when a 20 MHz spectrum is used. A device using 10 MHz spectrum with TDD scheme typically has peak physical layer data rate of about 25 Mbit/s in downlink and 6.7 Mbit/s for uplink. WiMAX supports adaptive modulation and coding (AMC) wherein modulation and coding schemes can be changed as per the user and the frame, based on channel

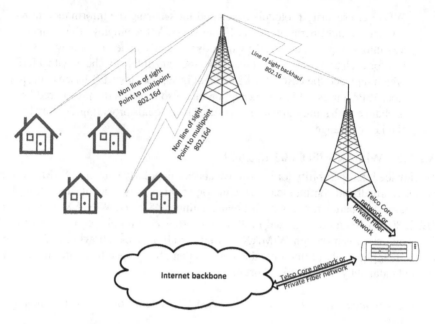

FIGURE 3.20   WiMAX network.

TABLE 3.6
IEEE Standards and Specifications for WiMAX

Design/Implemented	Standard	Usage	Throughput	Range	Frequency
1999/2004	802.16d	MAN/Fixed	75 Mbit/s	50 km	2–11 GHz
2001/2005	802.16a	MAN/Fixed	75 Mbit/s	50 km	2–11 GHz
2003/2006	802.16e	MAN/Mobile	75 Mbit/s	50 km	2–11 GHz
2006/2009	802.16m	MAN/Mobile	100 Mbit/s	50 km	2 & 11 GHz

conditions from a number of schemes that are supported. WiMAX at the link layer supports automatic re-transmission requests (ARQ) to achieve enhanced reliability and also supports both time division duplexing and frequency division duplexing, as well as a half-duplex FDD that enables a low-cost system implementation. WiMAX MAC has a scheduling algorithm for medium access in which the device has to compete only once for accessing the network after which an access slot is provided by the base station. The time slot always remains assigned to the device, but the slot can enlarge and contract. WiMAX supports strong security features using AES and offers a very robust authentication architecture based on extensible authentication protocol (EAP), which permits a variety of user credentials for accessing the network including username and password, digital certificates, and smart cards. WiMAX forum also has defined a reference network architecture where end-to-end services are delivered over an IP architecture with end-to-end transport, quality of service (QoS), session management, security and mobility, all implemented based on IP protocols.

3.2.2.4 ZigBee – IEEE 802.15.4-2003

ZigBee is IEEE 802.15.4-based wireless communication technology used to create personal area networks with low power and low bandwidth requirements providing a range of 10–100 m and a defined rate of 250 kbit/s depending on power output and channel characteristics. ZigBee modems available in the market can transmit data over larger distances but usually ZigBee uses mesh network to transmit data over larger distances in which intermediate devices are used to reach farther ones. ZigBee was conceived in 1998, standardized in 2003 and revised in 2006.

1. *Components and Techniques*: A device belonging to ZigBee network has two modes of operation – namely, full function device (FFD) and reduced function device (RFD). The FFD can carry out all the tasks specified in IEEE 802.15.4 standard whereas RFD can perform only a limited number of tasks. RFD normally handles only application task and does not take part in any network configuration and management thereby saving energy. ZigBee devices are of three kinds (Figure 3.21):

   a. *ZigBee coordinator* (ZC): ZigBee coordinator forms the root of a ZigBee network (only one ZigBee coordinator exists in a network), typically starts the network and acts as a master for all other members. It stores all the information about the network and manages the operation of the entire network; it is sometimes referred to as a personal area network (PAN) coordinator and is an FFD.

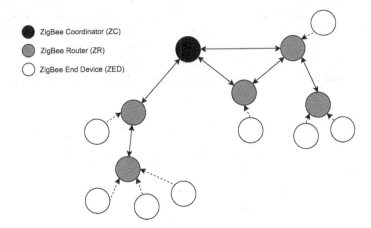

FIGURE 3.21 ZigBee network.

b. *ZigBee router* (ZR): A ZigBee router is an FFD that has the primary function of extending the network. Hence, like the coordinator, a router allows other devices to connect and communicate with it and the router also communicates with the coordinator. It can also run an application function and act as an intermediate device to facilitate passing of data between other devices.

c. *ZigBee end device* (ZED): A ZigBee end device is an RFD that only implements application tasks and functionality to talk to its parent node. It cannot do any network management activity and hence, can remain asleep when no application task is to be executed thereby giving long battery life. A ZED requires the least amount of memory and computational capabilities unless demanded by any application and is less expensive compared to a ZR or ZC. The ZEDs cannot communicate to each other directly, but they can exchange data via a ZR or ZC as they are RFDs.

ZigBee protocol supports various network topologies and the most commonly used topologies are star, mesh and cluster tree (Figure 3.22). The network in star topology has one coordinator that initiates and manages the devices in the network. All other devices in this topology are end devices that directly communicate with the coordinator. In mesh and tree topologies, the network has several routers in between the coordinator and the end devices. A router in mesh is either connected to other routers and coordinator or to many other routers to create multiple redundant paths so as to cater to network functionality even in case of node failure. In a cluster tree network, each cluster consists of one coordinator with leaf nodes as router and end devices; all these routers are connected to parent coordinator which initiates the entire network.

2. *Standards and Specification*: The physical and MAC layer is defined in IEEE 802.15.4 and ZigBee alliance has made specifications of network layer, application layer, security layer, Zigbee device objects (ZDOs) and application objects on top of all (Figure 3.23). The physical layer defines the operation of the device with respect to its communication capabilities in terms of sensitivity, power, modulation and demodulation and transmission rate. A ZigBee device supports 16 channels at 2.4 GHz ISM band at a data rate of 250 kbit/s, 10 channels at 915 MHz ISM band at 40 kbit/s and 1 channel at 868 MHz with a data rate of 20 kbit/s. MAC layer manages

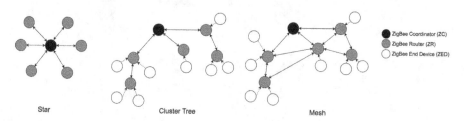

FIGURE 3.22 ZigBee network topologies.

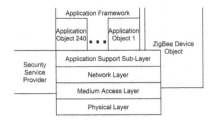

	Band	Coverage	Data rate	Number of channels
2.4 GHz	ISM	Worlwide	250 kbps	11-26
868 MHz		Europe	20 kbps	6
915 MHz	ISM	Americas	40 kbps	1-10

FIGURE 3.23 ZigBee protocol stack.

the medium access and interactions among the neighboring devices and includes services for retransmission, acknowledgement and collision management. MAC layer of IEEE 802.15.4 supports beacon enabled and non-beacon enabled operation. In non-beacon type an un-slotted CSMA/CA mechanism is used for channel access where the ZigBee routers have to be continuously active waiting for incoming signal during the entire communication window. In beacon enabled networks, the ZigBee routers transmit periodic beacons to synchronize the operation of all network nodes. Nodes wake up on beacons' request for a slot to transmit data. A node sleeps till the slot arrives, then wakes up and transmits data; else, it competes using slotted CSMA/CA and transmits the data. Once transmission is done, the node sleeps till the arrival of the next beacon and thus increases the battery life. Beacon intervals depend on data rate and long beacon intervals are preferred for low duty cycle operation; but this increases the delay and so precise timing based on trade-off is needed. Network layer implements all network related operations such as network setup, management network topologies, connecting, disconnecting, address allocation and deallocation, connection and disconnection of end devices, and routing of data packets. The routing protocol used is ad-hoc on-demand distance vector (AODV) that broadcasts a route request to all neighbors which the nodes further forward if required. The route reply following the lowest cost path is used to update the routing table with required entries. Application layer is the topmost layer of the protocol stack that houses ZDO and application objects. The application layer provides all services to user applications and manages device profiles. The ZDO provides features for device management and security policies and secure link management. The application support sublayer enables the services and interfaces, necessary for all protocols and layers, for data management. ZigBee implements strong security features, which are built on the basic security framework defined in IEEE 802.15.4. The security features provide access control via authentication, data integrity via encryption using symmetric key cryptography and message integrity checks to ensure integrity of transmitted messages. ZigBee uses three different types of keys – namely, master key, link key and network key – and the key establishment is done via pre-installation, key transportation and symmetric-key key establishment (SKKE) protocol.

3.2.2.5 Bluetooth – IEEE 802.15.1-2005

Bluetooth is a wireless technology standard managed by the Bluetooth's special interest group and standardized by IEEE as 802.15.1 used for data exchange among devices over short distances in the 2.4 GHz ISM band. It is considered as a personal area network technology. Bluetooth allows communication between two devices in a master-slave mode.

1. *Components and Techniques*: A master Bluetooth device can form a Bluetooth network called *piconet* for communication with a maximum of seven other devices. A slave can switch role and become the master any time enabling provisions for creating more piconets. Two or more piconets connected together can form a *scatternet,* in which some devices simultaneously act as the master in one piconet and the slave in another (Figure 3.24). Normally, data transfer happens between the master and another device. The master chooses a slave device to address and switches from one slave device to another in a round robin fashion when data exchange is required with all slaves. A Bluetooth device in discoverable mode always broadcasts the device name, class, list of services and some technical information; it usually needs pairing or acceptance to get connected to a device. A Bluetooth device has a unique 48-bit address and pairing can be achieved via secure simple pairing.

2. *Standards and Specification*: Bluetooth uses frequency hopping spread spectrum (FSSS) in its physical layer and transmits data as packets on one of 79 channels (around 40 for Bluetooth low energy with 2 MHz spacing) having a bandwidth of 1 MHz. With adaptive frequency hopping, Bluetooth usually performs 1600 hops per second. Originally, Gaussian frequency

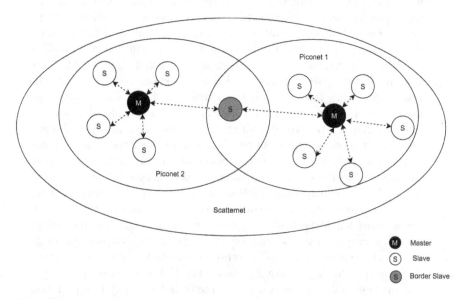

FIGURE 3.24   Bluetooth network.

TABLE 3.7
Bluetooth Specifications

Features	Bluetooth 5	Bluetooth 4.2 (BLE)	Bluetooth Classic
Maximum data rate	2 Mbit/s	1 Mbit/s	1 Mbit/s
Range (Outdoor)	10–40 m	40–240 m	10–40 m
Power	Lowest	Low	High
Channel spacing	2 GHz	2 GHz	1 GHz

shift keying (GFSK) modulation was used in Bluetooth operating in basic rate mode to achieve bit rate of 1 Mbit/s. From Bluetooth 2.0, differential quadrature phase-shift keying (π/4-DQPSK) and differential phase-shift keying (8-DPSK) are being used in enhanced data rate mode, each giving 2 and 3 Mbit/s respectively (Table 3.7). Error handling in Bluetooth includes forward error correction and retransmission under automatic repeat request scheme. A *Link Manager* establishes, authenticates and configures the links for connection between devices in a Bluetooth network via the management protocol called as link manager protocol (LMP). LMP sends a number of protocol data units (PDU) from one device to another and uses the services included in the *Link Controller*. The host controller interface sets up an access layer and enables automatic discovery of other Bluetooth devices that are available within the coverage range. The *Logical Link Control and Adaptation Protocol* (L2CAP) helps in making the right logical connection between two devices and also does segmentation and reassembly of packets. The *Service Discovery Protocol* helps to discover services and the associated parameters offered by other devices and identified by a universally unique identifier (UUID). The *Radio Frequency Communication* (RFCOMM) generates virtual serial data stream over the Bluetooth baseband layer to the user. The *Bluetooth Network Encapsulation Protocol* provides networking capabilities and transfers another protocol's data via L2CAP channel. Security in Bluetooth is achieved via authentication, authorization and encryption. It offers various security modes with devices treated as trusted or untrusted ones; the services considered as open to all are: (i) That needs authentication only, (ii) that needs both authentication and authorization, and (iii) that has encryption (Figure 3.25).

3.2.2.6 Z-Wave

Z-Wave, developed by Zensys in 1999, is a wireless communications technology that operates in the ISM bands in the 850–930 MHz frequency range and is primarily used for home automation applications with a maximum range of about 100 m. Z-Wave system renders application layer interoperability by allowing message passing. Therefore, control of devices of Z-wave network via other networks and

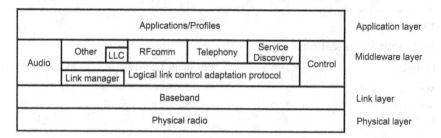

FIGURE 3.25 Bluetooth protocol stack.

applications – by means of a gateway or a primary control device serving as the main network controller and link to the outside network – is possible. Z-wave supports mesh network topology of devices to communicate from appliance to appliance enabling application needs and defines three types of devices.

1. *Components and Techniques*: The device that creates a new Z-wave network acts as the primary controller and is a master of the network which can add and remove nodes in the network, issue node IDs, track complete topology and routing tables and issue control commands to all other members of the network, thereby managing the network operation fully. The secondary controllers are those nodes that receive routing table from the primary controller and perform routing, typically acting as a repeater. The third and the last type of device in a Z-Wave network is the end device that receives the commands and acts accordingly; it cannot do any direct communication with any other end device. The master controller can make any end device a secondary controller by using commands (Figure 3.26).

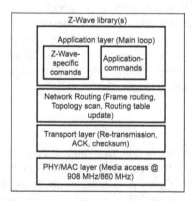 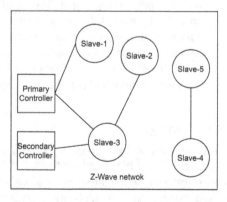

FIGURE 3.26 Z-Wave protocol stack and network structure.

2. *Standards and Specification*: Like all other protocols, the physical and MAC layer together controls the access to RF media in Z-Wave too. The physical layer uses frequency shift keying (FSK) modulation with Manchester channel encoding for 9.6 kbit/s transfer rate operation and non-return to zero (NRZ) encoding for 40 kbit/s operation. An improvement in data rate facilitating up to 100 kbit/s is achieved by using GFSK with NRZ encoding. The MAC layer uses CSMA/CA for medium access with reduced collision. Node and network identifier management is also done at this layer. Every Z-Wave network has a unique 32-bit identifier called the *Home ID* and an 8-bit identifier to address all the devices. Transport layer manages the data transfer between devices and handles retransmission of data, checksum calculation and acknowledgement. It supports four frame types – namely, singlecast, multicast, broadcast, and acknowledgement. Network layer is responsible for routing of data packets in the network, scanning of network topology and maintaining the routing table. Source-based mesh routing protocol is implemented in Z-Wave and all nodes in the network have the capability to automatically route the data from one node to the next till the destination is reached. Wireless mesh networking with source routing allows any node of the network to talk to the adjacent nodes directly or indirectly, even directly to nodes that are not within the range via another node that is within the range of both the nodes, as the source nodes specify the list of intermediate nodes between the source and the destination in the packet header. Application layer is responsible for decoding and execution of commands, besides providing interoperable features that allow devices to share messages and associated hardware/software to work together. Z-Wave uses AES symmetric block cipher algorithm using 128-bit key for data authentication by encryption and it has a custom key establishment protocol (Table 3.8).

TABLE 3.8
Z-Wave Specification

Parameter	Specification
Frequency range	868.42 MHz in Europe, 908.42 MHz in US
Data rate	9.6–100 kbit/s
Number of nodes	232
Medium access	CSMA/CA
Modulation	FSK (for 9.6 and 40 kbit/s), GFSK with BT = 0.6 (for 100 kbit/s)
Coding	Manchester encoding (for 9.6 kbit/s), NRZ encoding (for 40 and 100 kbit/s)
Range	Indoor – 30 m, outdoor – 100 m

3.2.2.7   Wireless HART

Wireless HART, released in 2007, is a secure and TDMA-based wireless technology developed to meet the requirements of process field device networks and is based on the highway addressable remote transducer protocol (HART). The technology uses a time synchronized, self-organizing, and self-healing mesh network architecture and operates in the 2.4 GHz ISM band using IEEE 802.15.4 radios.

1. *Components and Techniques*: A typical wireless HART network defines two types of devices – namely, *field devices* member nodes of the network and a *network manager* that is responsible for network management and scheduling. Additionally, gateway nodes, diagnostic devices and servers can be part of the network (Figure 3.27).

2. *Standards and Specification*: Wireless HART protocol stack has a layered architecture with physical layer at the bottom, data link layer as the second layer from the bottom, network layer as the third, transport layer on top of it and application layer as the topmost layer. In addition to the protocol stack, wireless HART defines a central network manager to handle routing and mediate the schedule for communication. Wireless HART protocol has IEEE 802.15.4-2006 as its physical layer and defines its own time synchronized MAC layer on top of it. The physical layer uses DSSS for modulation and provides a data rate of 250 kbit/s and defines the

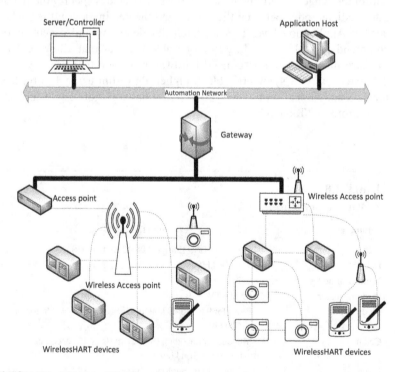

FIGURE 3.27   Wireless HART network.

radio characteristics, such as the signaling method, signal strength, and device sensitivity. Wireless HART MAC has strict 10 ms time slot and uses TDMA for medium access with a periodic superframe. Channel blacklisting is a process in which channels with consistent interferences that result in poor packet delivery are blacklisted and the network can avoid the use of such channels. Also, each node maintains an active channel table so as to help the channel hopping so that blacklisted channels are not selected while hopping. The network layer implements self-organizing and self-healing routing strategy for the mesh network and a central network manager maintains the up-to-date routes for scheduling effective routing in the network. The network layer supports both graph-based routing and source-based routing. In graph-based routing, the network manager explicitly creates path for each graph and is provided to each individual network device. Any device that wants to transmit a data packet mentions the graph ID in the packet header and routes the packet to the destination node. In source routing, the source node specifies the devices that must act as repeaters or intermediate nodes for a packet from the source to destination node in the packet header itself. The network layer and the transport layer of wireless HART together ensure secure and reliable end-to-end communication. The application layer implements measures for passing the message content, extracting the command number, executing the specified command, generating responses, and reporting status. MAC and network layer together implement security features in wireless HART. Message integrity check and encryption using 128 bit AES is provided at the MAC layer and public, session, join and network keys are implemented in the network layer to provide confidentiality and data integrity (Figure 3.28).

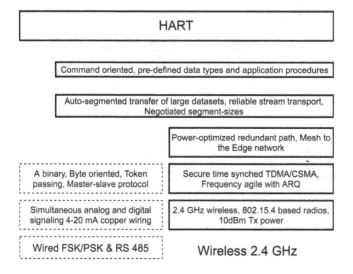

FIGURE 3.28 Wireless HART protocol stack.

3.2.2.8 Insteon

Insteon launched in 2005 by Smartlabs is a communication technology solution that enables multiple devices to communicate using power line communication, radio frequency (RF) communication or both for automation applications. Member devices in Insteon are referred to as peers and all peers can transmit, receive and repeat messages independently without the intervention of any master device; this uses a dual mesh networking topology to have an unsupervised message passing without any routing table.

1. *Components and Techniques*: Insteon has three type of devices in the network: The first can communicate using RF only, the second can communicate using power line only and the third can use both power line and RF. All three types can act as controllers and responders. Controllers are devices that send signals to other devices of the network based on manual configurations or automated events, like a programmable or smart network device. Responders receive the commands sent by controllers and act accordingly. They are typically devices that control the appliances. The responder nodes are application-specific and have to be selected and included in the network particularly and cannot be generic as they have to interact with the specific systems at a home. Every Insteon device acts as a repeater and will do this automatically as soon as they are powered up (Figure 3.29).
2. *Standards and Specification*: Insteon powerline messaging uses binary phase shift keying (BPSK) and a frequency of 131.65 kHz. The standard

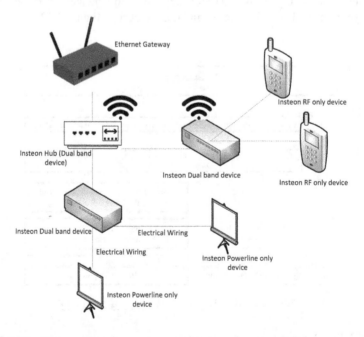

FIGURE 3.29 Insteon network.

message has 5 packets and has a size of 120 bits and extended message has 11 packets with 264 bits. Insteon RF messaging uses frequency-shift keying at 869.85/915/921 MHz; the standard message has 112 bits and extended message has 224 bits. Error detection and correction is done on each message received by an Insteon device and is then retransmitted so as to have better reliability. The retransmissions are done simultaneously so that it is synchronous to the powerline frequency. All messages have a two-bit hop count field which normally is set to 3 by the source node and reduced every time a member node of the network retransmits the message. Insteon being a peer to peer mesh network, the devices function without central controllers and routing tables, but any computing device can be made to act as a central controller to implement functions such as scheduling, event handling and problem reporting via email or text messaging with a compatible Insteon modem connected to the computing device. Insteon power line protocol has one-way interoperability with the X10 protocol such that Insteon devices can listen and talk to X10 devices, but X10 devices cannot talk to Insteon devices. Insteon network security is maintained via linking control where a link can be created to a device only if physical possession is there or the unique device ID is known and rolling codes for encryption in the case of extended messages (Table 3.9).

3.2.2.9   Wireless M-Bus

The wireless M-Bus is a wireless communication protocol for communication system and remote reading of meters based on the M-Bus or Meter-Bus protocol, which is a European standard defined under EN 13757 and is defined under EN 13757-4.

1. *Components and Techniques*: The wireless M-Bus protocol allows meters, data concentrators and servers as member nodes of the network (Figure 3.30). The meters are configured to transmit the consumption data to the data concentrators and the data concentrators forward the same configured in proper

---

TABLE 3.9
Insteon Specification

Parameter	Specification
Frequency	915/869.85/921 MHz (RF PHY), 131.65 kHz (Power line PHY)
Modulation	FSK – RF PHY, BPSK – Power line PHY
Range (RF PHY)	150 feet unobstructed LOS
Message type	10 bytes (standard), 24 bytes (extended)
Devices supported	Pre-assigned address (24 bit) to support 16777216 devices
Security	Physical access, address masking, encryption

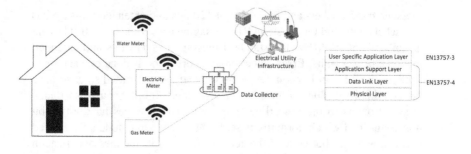

FIGURE 3.30 Wireless M-Bus network and protocol stack.

format to the server at utility or control centers. The data concentrators directly or servers via the data concentrators can command the meters to obtain the data beside the configured format and time.

2. *Standards and Specification*: The physical layer for the wireless M-Bus provides transmission at three different frequencies of 169, 433 and 868 MHz ISM band for wireless data exchange in one-way or two-way communication, with data rates ranging from 4.8 to 100 kbit/s as desired by the various modes of operation, and involves modulation types as Manchester, NRZ and 3-out-of-6 encoding. The data link layer supports two types of frame formats – one with preamble, payload bytes and post-amble, and the second without post-amble. Preamble has header and synchronization information and post-amble is a short bit sequence used only in selected modes. The payload data is encrypted with AES 128 bit cipher block chaining encryption. The application layer defines various profiles that help identify devices from different vendors that work well with wireless M-Bus system in an interoperable fashion.

3.2.2.10   6LoWPAN

IPv6 over low power wireless personal area networks (6LoWPAN), defined in RFC 6282 by the Internet engineering task force (IETF), is a wireless technology that allows transmission and reception of IPv6 packets over low data rate IEEE 802.15.4 using encapsulation and header compression techniques. Even when the inherent nature of IP and 802.15.4 are different, 6LoWPAN intends to bring the data delivery and connectivity capabilities of LAN/MAN/WAN as well as sensing and communication capabilities of PAN in the wireless domain. 6LoWPAN technology was devised with an idea to make IP capabilities available even in the smallest applications having low power and feature-restricted devices so as to make all of these part of the Internet of Things (IoT). Another interesting aspect in 6LoWPAN development is the effort to make it interoperate with Sub-1 GHz low-power RF, Bluetooth, PLCC and low-power Wi-Fi, even though the initial goal in conceiving it was to provide IP capabilities to IEEE 802.15.4-based, low power and low data rate wireless networks in the 2.4 GHz ISM band.

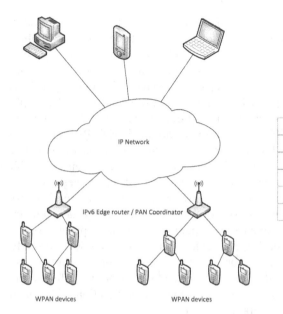

| Application Layer |
| Transport Layer (TCP/UDP) |
| Network Layer (IPv6) |
| 6LoWPAN Adaptation Layer |
| IEEE 802.15.4 Link Layer |
| IEEE 802.15.4 Physical Layer |

FIGURE 3.31 6LoWPAN network & protocol stack.

1. *Components and Techniques*: A typical 6LoWPAN network along with an IP network is shown in Figure 3.31. The coordinator of the 6LoWPAN network acts a router in the IP network operating on edge. This edge router handles the data between 6LoWPAN network and the IP network, local data exchange amongst devices in the 6LoWPAN and creation and maintenance of 6LoWPAN network. Besides the edge router, the 6LoWPAN network has normal routers and host devices. Routers route packets inside the 6LoWPAN network and the host is an end device that acts in the physical world outside with sensing or actuation.

2. *Standards and Specification*: 6LoWPAN uses the IEEE 802.15.4 protocols at the physical and MAC layers and a 6LoWPAN adaptation layer provides adaptation from IPv6 to IEEE 802.15.4 above it. A physical layer amendment IEEE 802.15.4g is used in 6LoWPAN that provides an extra range of Sub-1 GHz frequency band. Similarly, IEEE 802.15.4e is an amendment in the MAC layer that enables enhancements such as time slotted channel hopping (TSCH) and coordinated sampled listening (CSL), both aiding in reducing the power consumption and making a more robust interface. The data link layer provides a reliable link between two directly connected nodes, as detection and correction of errors, accessing medium using CSMA/CA and data framing happen at this layer. The 6LoWPAN adaptation layer provides header compression, fragmentation and reassembly, and stateless auto configuration services to facilitate transmission of IPv6 packets over low power and lossy networks (LLNs) such as IEEE 802.15.4. The network layer provides

addresses to all devices and manages routing using IP. The transport layer creates and manages communication sessions between multiple applications on each device and supports both TCP, the connection-based protocol, and UDP, a lower overhead and connectionless protocol. Transport layer security (TLS) running on top of TCP and datagram transport layer security (DTLS) on top of UDP ensure secure transport layer. The application layer takes care of data formatting using optimal schemes for various applications. Hypertext transfer protocol (HTTP) over TCP is a common IP choice, but has large overhead. Constrained application protocol (COAP) runs over UDP and defines retransmissions, verifiable and non-verifiable messages, support for sleep state in devices, blocking data transfers, data subscription and resource discovery. Another protocol in the application layer of 6LoWPAN is message queue telemetry transport (MQTT), an open-source publish/subscribe type of protocol running over TCP. The MQTT devices do not communicate directly, instead it communicates via brokers using wild cards to access topics that refer to data. Each device publishes and subscribes to different topics. Upon an update on the subscribed topics a notification followed by the data will be received via the broker.

3.2.2.11 Wavenis

Wavenis is a wireless communication technology developed by Coronis in 2000 that supports communication of ultra low-power and long-range devices and operates with 9.6 kbit/s in 433/868 MHz ISM band and 19.2 kbit/s in 915 MHz ISM band. Wavenis consists of an RF transceiver and a protocol stack and the combination is presented as a software and hardware solution for secure and reliable wireless communications over long range with minimal power consumption.

1. *Standards and Specification*: Wavenis uses frequency hopping spread spectrum along with Gaussian frequency shift keying (GFSK) for signal transmission and provides a line of sight range up to 1 km with a maximum supported data rate of 100 kbit/s. Different techniques such as data interleaving, forward error correction and automatic frequency control (AFC) ensure power saving, high performance and robust networking. MAC layer presents synchronized and non-synchronous medium access mechanisms. A mixed CSMA/TDMA approach is used in synchronized scheme for medium access wherein a pseudo random time slot, based on node address, is chosen for transmission, as per TDMA scheme, after *carrier sense* at the beginning of the slot. The CSMA/CA is used for medium access in non-synchronized scheme. The logical link control handles flow and error control by acknowledgements for frames and windows. Wavenis defines only one type of device and supports tree and star network topologies. The root of the tree as well as the central node in star may act as a sink or gateway. A new device can join the network by broadcasting a request with

sufficient QoS value based on received signal strength indicator (RSSI), remaining battery capacity and number of devices already connected. The response from the best possible match is used by an incoming node to join a network and the same parameters are used in the self-discovery protocol (SDP) for dynamic routing table updation. The host to controller interface provides simple APIs for the application functions to access the Wavenis services (Figure 3.32 and Table 3.10).

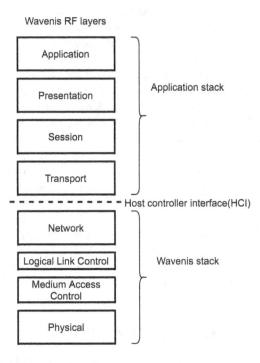

FIGURE 3.32   Wavenis protocol stack.

TABLE 3.10
Wavenis Specification

Parameter	Specification
Frequency range	868 MHz, 915 MHz, 433 MHz
Data rate	4.8–100 kbit/s
Physical layer	GFSK + FHSS
Coverage range	1 km (Indoor)
Security	128 bit AES
Network topology	Tree and star

3.2.2.12 Cognitive Radio – IEEE 802.22-2015

A dynamically programmable and configurable device that can select the best available wireless radio channel in the vicinity so as to have minimal interference and congestion is a cognitive radio (CR). A device like this automatically discovers all available channels in wireless spectrum in its vicinity, and then changes its transmission or reception parameters so as to allow more simultaneous wireless communications in the given spectrum at a given location. Depending on transmission and reception parameters, two main types of cognitive radio are defined – namely, a *Full Cognitive Radio* which considers all possible parameters observable by a wireless node and a *Spectrum Sensing Cognitive Radio*, that considers only one radio frequency spectrum. Further, the spectrum sensing cognitive radio has so many types based on which all parts of the spectrum are made available for use like *licensed-band cognitive radio, unlicensed-band cognitive radio, spectrum mobility, spectrum sharing, sensing-based spectrum sharing*, and *database-enabled spectrum sharing*.

Cognitive radio was initially brought as an extension to the software defined radio concept that led to development of full cognitive radio; but, later most of the research in this area focused on spectrum sensing cognitive radio. The biggest challenge in spectrum sensing cognitive radio is the design and development of high-quality spectrum sensing devices and algorithms. The main functions of cognitive radios are *power control, spectrum sensing*, and *spectrum management*. Power control is normally used in spectrum sharing cognitive radios to maximize the capacity of users considering interference power constraints. Spectrum sensing is the process of detecting unused spectrum and sharing it without adversely affecting the users. Spectrum management involves spectrum analysis and decision making to decide on the best spectrum band for meeting the user communication requirements, while not creating undue interference to other users.

3.3 COMMUNICATION REQUIREMENTS IN SMART GRID

Smart grid is an umbrella term that envisions every possible smart technique that can be incorporated into various sectors of operational aspects and application services in electric power system so as to realize efficient, reliable, affordable and quality electric power delivery. Smart grid will thus have many new applications for all stakeholders and these applications will include many interrelated systems. One key technological area that is invariably required in almost all automation that smart grid brings forth is ICT that provides bidirectional communication capability to all devices involved. A good understanding on the ever-evolving communication requirements, with each new application, of all smart grid stakeholders is essential for smooth design, development and integration of ICT to power system to make it smart (Figure 3.33).

There exists many detailed studies and reports that give a clear indication as to how the communication needs are being satisfied with in the existing grid with present automation level and what the expected network requirements would turn out to be with adoption of more technologies and applications in realizing a full-fledged smart grid. Almost all of the existing network and communication

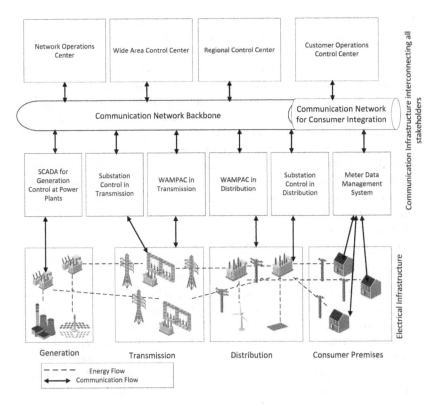

FIGURE 3.33 ICT infrastructure for smart grid.

technologies can be of use to satisfy some application in one or the other sector or service in smart grid. Various smart grid ICT networks that can be developed using these technologies to satisfy the smart grid application requirements include building area networks (BAN), home area networks (HAN), field area networks (FAN), neighborhood area networks (NAN), WAN, backbone networks and supervisory control and data acquisition (SCADA) networks. These smart grid ICT networks provide data transfer facility to satisfy needs of automation of substations and generating stations including distributed generation, wide area situational awareness, protection and control in transmission and distribution sectors, and demand side applications. It is very important to keep in mind that communication requirements for the existing assets and their near-term evolutions may be easy to be formulated, but future communication needs may be difficult to be quantified due to the rapid rate at which evolution is happening in smart grid technologies and applications.

Automation of any power system application can be composed of three generic systems – namely, Master/Central Station, the end devices commonly grouped as remote terminal units (RTU) and the communication network providing interconnection among all associated devices.

Master stations: A power system is so vastly interconnected that action taken at any one point impacts the operation of the entire system and hence, decision-making cannot be done in small pockets. Any decision that has to be made for the operation and control of power system can be done only with a large amount of distributed data considering the state of the system. Hence, the role of master stations is highly important, and it is very much evident from the traditional operation of power systems that the central or master stations of various applications interoperate and share data for fruitful management of power system operation. The early SCADA systems were installed for automation of high voltage substations and power plants in 1920s so as to monitor and control their operation from an off-site control room. As individual utilities started to interconnect for commercial and operational reasons, precise generation control became a need and systems for monitoring and control of generator output, tie line power flow and frequency were in place from 1930s at various levels of operation and control centers. With enhanced analog computers and offline manual calculations, various applications and services under the ambit of energy management systems (EMS), like economic dispatch and automatic generation control for generation scheduling and tie line power transfer, were in place in 1950s. Digital computers and software systems started replacing the analog EMS including offline analysis of transmission system models in late 1960s. The computing resources and software services started to undergo regular upgrades and the concept of master systems gave way to hierarchical control systems to meet the operational demands and the new real-time applications. Different horizontal and vertical layers of control centers were created in the hierarchical approach so as to control various applications catering to the needs of different stakeholders of smart grid. The bottom layers provide monitoring and control of various consumer side applications, the middle layers do it on the transmission and distribution grids and the higher layers address the needs of generation plants, energy trading and higher-level energy transactions. This reduced the complexity of centralized EMS in terms of computational resources, database and amount of information input by introducing distributed management. Many utilities have started deploying such distributed control systems with area wise transmission and distribution control centers and also regional distribution management systems (DMS) which communicate with local distribution substations and devices along the distribution feeders. Some of these master stations are located in the substations acting either independently or on instructions from network operation centers. Different utilities use different communication technologies including wired, wireless and even hybrid systems to facilitate communication networks for such control centers managed by independent system operators (ISOs) or regional transmission system operators (RTOs) providing real-time monitoring, faster decision making, better determination of system state and improved response time for control actions and restoration.

End terminal units: Traditionally the end terminal devices are used for data collection and actuation and consist of terminal provisions for the interconnection to the power system. These are also referred to as remote terminal units (RTU) in SCADA systems. The RTUs have the capability to communicate digital values representing the state of various power system devices and analog values which are various measurements made in the power system as series of binary values to the master station for display, analysis and storage. That the RTU and the sub modules are addressable, enabled the operators or master station systems to remotely command and control these. By the second half of 1970s rugged and robust microprocessors reduced the hardware complexity of the RTU and also increased its capabilities. In the 1980s, microprocessors were included in various power system devices – such as, protective relays, meters and controllers – in order to accommodate more functionalities. Because of their inclusion these devices were able to interface directly with the RTU via communication ports. It unfolded a new era of opportunities in power systems. With many applications and services getting enabled this way, the demand for more functionalities and features too grew, paving way for development and grouping of all such capable end terminal devices in power system as intelligent electronic device (IED), a term coined and defined by IEEE Power and Energy Society (PES) Substations Committee. The capabilities of these devices included time keeping, computational capability to run more complex and powerful protocols, individual point numbering, local logging and time tagging of events, higher communication speeds, multiple communication ports, better local analysis and numerous other functions. By 1990s, many utilities started installing IEDs on their distribution feeders, in addition to the transmission feeders and the various station automations, with communication capability for data exchange with the local substation RTU and even with the network operations center. That made IED the preferred interface between the power system and the RTU. Currently, IEDs exist in various forms and classes in very large numbers on many transmission and distribution feeders. These regularly share the analog measurements and status data with the SCADA master and various levels of control centers, thus providing monitoring and control capability to the system operators.

Communications systems: The first phase of utility monitoring and control system worked with leased telephone lines to interconnect various stations, plants and control centers. The performance and capabilities of the system evolved better with high-speed switching and automatic fault recovery features. Another preferred choice of many utilities, particularly among substation and control centers, was PLCC. Supporting both voice and data transfer, it seemed to be a good choice with a direct link between the two substations or control centers. Microwave and satellite communications were also used by some utilities for sparsely populated large geographical areas to handle big volumes of information over long distance. Fiber optic cable became the popular choice in the new century within substations and plants as well as for WAN. Small substations and distribution sector applications started

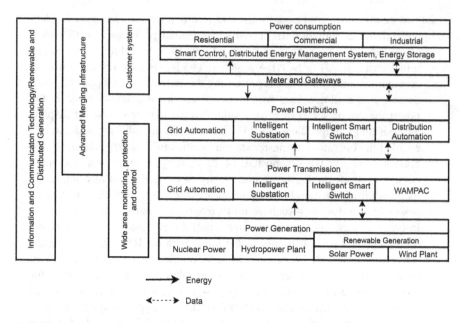

FIGURE 3.34 Overview of sector wise smart applications in smart grid.

using the licensed 900 MHz point to multipoint systems and the unlicensed 900 MHz mesh radio systems among the various communication technology mix available. All these technologies, especially in the distribution sector and further at the consumer side, have own advantages and disadvantages and there is no final winner as many utilities and applications use one or a combination of many technologies to meet the requirements (Figure 3.34).

Protocols & standards: Protocol is the glue that allows smoother operation of all units of any system, particularly a networked system, facilitating multiple units to operate together for satisfying the need of a common application. The electric utility industry and services started with hardware-based devices from a few manufacturers and with the digitalization of the systems more vendors and more products came along. Eventually hardware solutions gave way to hardware-software products which made inter operation and network wide integrated operation challenging because of the wide variety of features and functionalities brought in by all these devices. The major suppliers of devices for power systems solved part of the problem by documenting their protocol and allowing customers to use it for inter operation. When IEDs and similar smart devices at all sectors of grid began to be marketed, the number of protocols almost exploded exponentially with each new vendor defining a protocol for their device; some even defining a new protocol for each new model. This provided a very hard time for system level vendors and utilities as they had to find means to integrate and interact with all these devices where many of these protocols were proprietary.

The utilities, vendors, research institutions and other stakeholders formed consortiums to have a solution to this problem and organizations like IEEE and International Electrotechnical Commission (IEC) lead the standardization activities. Standards were developed which clearly spelled out what a device, service, solution and associated protocols should offer to meet the requirements of various applications spread across all the very wide spectrum of power system network and this stopped every device and service coming to the market with a new protocol. As the concept of smart grid is still evolving in terms of various services and applications at different sectors of power system, standards are still being defined and reviewed for many applications in power system and at the same time standards are finalized for a big group of applications.

Even when considered separately, ICT stands out among the many smart grid technologies as each and every smart grid application calls for it with varying requirements and it is expected to decide how the electricity network employs automation to bring in smartness. The smart grid in all likelihood will have many communication technologies, to satisfy the demands from multiple applications and performance review of various technologies will reveal where it may be used within the grid. Most sectors of smart grid will have multiple applications that will function using the same network; performance assessment of the communication technologies would need to consider together the various requirements of all such applications too. The following sub sections present the communication requirements of various applications in all these sectors.

3.3.1 COMMUNICATION REQUIREMENTS IN GENERATION

Any power generation utility aims to maximize profit through customer satisfaction and operational efficiency by 24 × 7 supply of reliable, quality and affordable electrical energy. This can be achieved in generation sector by efficient and smart mechanisms like SCADA or similar automated systems, irrespective of the type of power plant. The utilities adopt an extremely cautious approach in selection and implementation of computerized automation in power plants owing to the multitude of subsystems present in a power plant which in most cases are not the same as, but are similar to, the systems of another power plant of the same type.

As the source from which electrical energy is generated varies, the automation system and the methods of implementation of various associated applications fully vary for different types of power plants. Since the components and subsystems that constitute a power generation plant is specific to the type of source, some applications and systems may be the same or similar. Therefore, the operations of power plant are normally automated by installing sensors and actuators at all locations that need a measurement or actuation for the smart operation of the whole plant and interfacing these units with various data collectors and hierarchical control centers where data processing and decision making happen. Every plant will have multiple subsystems for providing various functionalities and some of these may have sensors, actuators and dedicated controls which are further interconnected, and data is made available at the central control room; data from the remaining subsystems

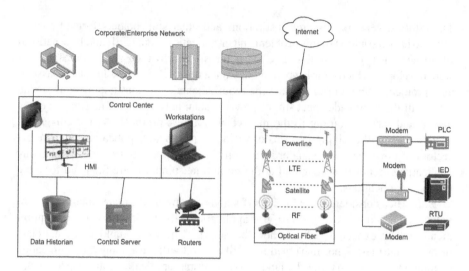

FIGURE 3.35   SCADA network architecture.

is fed to the control center directly or via data collectors. Some operations in the control center are decided based on information from outside the network like the amount of power generated, amount of power injected to the grid and number of reactive power transactions. Also, there is data to be shared with system operators, operational control centers and energy markets. So, in addition to providing power plant automation, the SCADA system must also facilitate communication with certain systems outside the plant (Figure 3.35).

The typical SCADA system for generic automation of a power plant has the following elements:

> *Remote terminal units/programmable logic controllers (PLC)*: Sensors and actuators for measurement and control are interfaced to RTUs, which are networked to the supervisory computer system or SCADA master. The RTUs are usually treated as intelligent end devices and often include processing and control capabilities so as to accomplish several logic operations. The particular sensors and actuators used will vary with the services rendered by the RTU and so the capabilities of various RTUs differ. The PLCs, in the context of a SCADA, work almost like RTUs as these are also interfaced to sensors and actuators that are to be automated and networked to the SCADA master. The PLCs used to have more advanced and focused digital control capabilities whereas RTUs were better telemetry devices. But with the evolution in embedded and PLC technologies, most RTUs and PLCs have similar capabilities now and are used interchangeably; a choice between the two is based equally on commercial and management considerations.
>
> *Communication infrastructure*: This provides the infrastructure for interconnecting the SCADA master to the RTUs and PLCs and uses industry

standards or proprietary protocols for data exchange. The interconnection can be achieved by means of wired and wireless networking. The choice between the two is application dependent; traditionally preference is for a wired system because of the inherent qualities of security, less interference and data loss and higher bandwidth. Critical systems will adopt redundancy by co-opting a secondary data highway as a hot standby.

Supervisory computer system/SCADA master: This forms the core of the SCADA system for automation. It gathers data of all processes and carries out processing, visualization and higher-level decision making, then sends control commands to the field-connected devices. It is a combination of integrated hardware and software, with a human machine interface (HMI) front end and data crunching and storage functions at the back end. The operator workstation itself forms the supervisory computer in small scale SCADA systems, and HMI is part of this computer. In large scale SCADA systems, the master station may include several servers for data acquisition, distributed software applications, communication and visualization with HMIs hosted on client workstations and the master level HMI system is accessible to the main operator or administrator. Critical SCADA systems will have the multiple servers often configured in dual redundant or hot standby modes via actual servers or virtualization – this enables real-time monitoring and control even in the case of a server malfunction or breakdown.

Human-machine interface: HMI is the visualization screen – or the operator window of the supervisory system – where the complete status of the plant is presented graphically mimicking the operation of the plant. The software interface provides supervisory personnel the ability to initiate control actions, as well as access alarm and event logging data. The HMI accesses real-time data from field devices via SCADA to provide live updates in the mimic diagrams with operational reports, alarm notifications, data displays and trending graphs. Mimic diagrams may be line graphics and schematic symbols or digital photographs to represent process elements or plant equipment overlain with animated symbols. Supervisory operation is achieved by means of the HMI in SCADA systems, with operators having provisions in HMI to issue commands using mouse pointers, keyboards and touch screens.

In addition to the HMI, the SCADA system may house software that can handle data processing – implementing various algorithms to achieve optimal operation of the generation plant under automation. These algorithms and software packages are often proprietary and specifically developed or customized for the plant. Another program that goes into the SCADA system is a database service that can accumulate time stamped data, events and alarms from the plant and can automatically populate graphical trends and reports in the HMI. This service is often referred to as a *historian* in a typical SCADA system and always works in connection with a data acquisition server.

The various generic functionalities of a SCADA system in a power generation plant are as follows:

- *Data acquisition or monitoring*: This involves collecting all necessary parameters including status and measurements of all the modules and systems involved in a power generation plant and is achieved by the combined operation of RTUs, SCADA masters, communication servers and data acquisition servers. Plant information system – an enterprise infrastructure for management of real-time data and events of the plant – is implemented using this functionality.
- *Status monitoring and process visualization*: This function deals with analyzing the data received through data acquisition and understanding the state of operation of the plant. This is done by presenting all relevant operational data in various graphical formats – like reports, charts and mimic diagrams – to operators and administrators who use the SCADA system. The presentation is achieved through data acquisition servers, historians, communication servers, data processing servers, SCADA masters and HMI. Online monitoring for real-time health and operational status is implemented using these functionalities.
- *Control configuration*: This allows system operators to configure the operation of various sensors and control devices based on local operational parameters or on the basis of information received from remote control centers and substations. This functionality allows operators to set thresholds for various events so that total system performance can always be maintained within limits. This is achieved by the use of HMI, SCADA master, communication server, and RTUs.
- *Process events and alarm management*: This involves the management of events and raising of alarms and notifications on events crossing the threshold configured, so that the operators take necessary steps to prevent the system from violating the set norms of operation at any instance. This functionality is achieved by RTUs, communication server, historian, and HMI.
- *Process operation management*: This involves carrying out various actions for operation management of the plant and includes generation control, scheduling, load balancing, reactive power control, active power control, energy management, etc., and is achieved by all constituent components of the SCADA system. This normally is not made possible by a single system but by the synchronized operation of various subsystems for individual activities. Most of these operations are critical and follow a real-time behavior.
- *Historical data logging*: This functionality is about logging all data, status or measured values and operational or administrative events, to a database facilitating access to information for further analysis and understanding. This is achieved by all units of the SCADA system with the historian and data acquisition server playing a major role. Data warehousing, a process for storing data for a longer time period, is implemented thorough this feature.
- *Communication management*: This function is to manage the data exchange among all modules of the SCADA system and also with the systems outside

the plant. The functionality includes facilitating a communication network for data exchange and providing security to the network. This ensures authorized access to the data and enough privacy and protection to the network from other private and public networks. The members of a SCADA network, at the same time, may also be part of other networks for different applications; hence ensuring that this does not create a vulnerability is also an important job.

- *Reporting and analysis*: This function is to analyze the data obtained from all systems so as to monitor the performance of the system through the *key performance indicators*. This is achieved by the data acquisition server, SCADA master and HMI.
- *Quality management*: This function ensures the availability of the SCADA system as well as the power plant throughout. This function considers the results of performance analysis and includes methods to alter the system operating conditions so as to maintain and even improve the quality from time to time. This function also provides provisions to analyze and find bad data amidst large volumes of measurement data and helps to remove errors in system operation.
- *Integration into business applications*: This function provides the information on generation to energy markets and also uses real-time information for financial analysis and operational improvement. Management information system, an enterprise service for business management, is implemented with the help of this function.

There will be more functionalities in a SCADA system that are very specific to the power plant of a generating station.

3.3.1.1 Communication Network for Power Generation Plant

Communication is, and will increasingly be, a very important tool that contributes to the optimal operation and maintenance of the power network and to its administrative processes. There exists a lot of communication technologies, both wired and wireless, that may be of use in meeting the needs of the automation system. So, while selecting the communication technology, care must be given on analyzing communication needs and requirements, providing a basis for a technical design. The interaction among various members of the communication network must take place in real time, while satisfying QoS. The main QoS parameters are delay at various levels, throughput, packet delivery ratio, overhead, number of devices supported, range, and cost. The limits for these parameters vary for different services of the automation network and many more specific performance parameters can be used in specific applications.

A communication network for a SCADA system that automates a power generation plant must provide the following two types of connectivity to meet the operational requirements (Figure 3.36):

LAN: The workstations, servers and other connected devices like RTUs, IEDs, and PLCs installed in the plant are to be interconnected to deliver the functions discussed above. All these devices are limited to a particular geographical

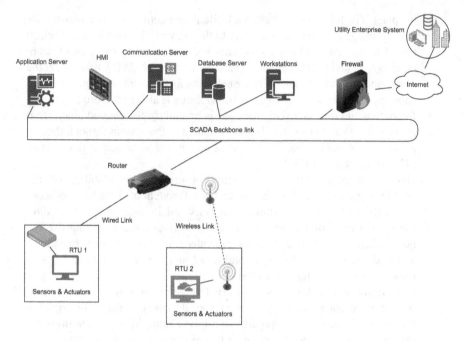

FIGURE 3.36 SCADA for generating station automation.

area of the power plant and interconnection of such devices creates a LAN. The interconnection links can be made using any communication technology; wired, wireless or even a heterogeneous network can be formed with different communication technologies at different links. The selection of a particular communication technology for any link or network depends on the match between functional and performance requirements and capabilities. The communication needs include message passing for monitoring, protection, status update, control and administrative applications. Protection applications have the most stringent requirements in terms of delay and reliability followed by control, monitoring, data logging and administrative applications. Throughput requirements are high for monitoring, data logging and administrative applications; status updates have medium throughput requirements and control and protection applications have low throughput requirements. Packet delivery ratio must be above average, overhead must be less and number of devices supported and range must not be too low.

WAN: Wide area network connections are to be created in the SCADA system for a power generation plant to provide interconnection between the devices of the SCADA network and the devices of a remote-control center, network operation center, substations or energy markets. Technically, this might not be a direct device-to-device connection, but interconnection of LANs at both places enabling an indirect connection so as to facilitate data exchange between the devices. This is achieved by access routers connected to a switch or hub on the LAN serving as a gateway to the WAN and forward the network traffic. As the applications over WAN are data exchanges, throughput and packet delivery ratio are the important performance metrics and both should be high. Delay must not be high and overhead must be low. WAN traffic usually travels via the Internet over public network or private network. Use of public networks is simpler and less costly, but has less security. Private networks use dedicated lines and therefore offer better security. Security can be improved in public networks by using virtual private networks (VPN) which create virtual "tunnels" through which data is communicated between two locations. Protecting access to the system by using passwords, authentication techniques and firewalls in switches and routers that connect to external networks are some mandatory requirements to ensure cyber security in this communication system.

3.3.2 COMMUNICATION REQUIREMENTS IN SUBSTATION

The substation automation is achieved using SCADA systems as in the case of generation power plants. The substation automation system (SAS) consists of following units (Figure 3.37):

IED: These end devices – which implement functionalities like data collection and actuation in a substation – have limited intelligence and are connected to master devices over a network.

Bay controller: This device acts as a master and controls the operation of all the devices connected to a single bay. It is connected to the SCADA master.

HMI: It is a workstation having an operator console with graphical interface for visualization, local control and system configuration.

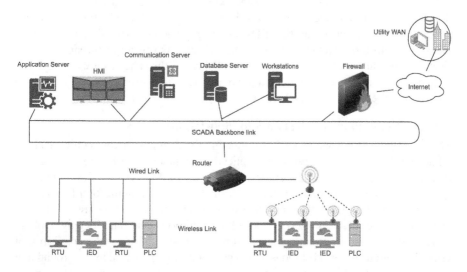

FIGURE 3.37 SCADA for substation automation.

Communication infrastructure: It provides facility for interconnection of various devices of the SCADA system to realize the operation of the system as
a whole.

SCADA master: It is the master control station that talks to bay controllers,
HMI devices and all other servers hosting various other services to manage
the complete system.

SCADA gateways: These devices provide interconnection of workstations, bay
controllers, SCADA master and various servers that host various services
of the system like data storage, security, communication services, other networks, etc.

Even though the applications and services of SCADA systems in substations differ from those in generating plants, the basic building blocks operate according to
their capabilities and do almost the same job. The operational constraints of these
modules vary from application to application and so do the ways these devices are
used to set up relevant architectures to meet the application requirements. A detailed
explanation of various basic modules that constitute a SCADA system is given in
Section 3.3.1.

Like in the case of SCADA in power plants the generic functionalities of SAS are
data acquisition or measurements, status monitoring & process visualization, control
configuration, process events and alarm management, process operation management, historical data logging, communication management, reporting and analysis,
quality management, integration into business applications, and time synchronization. Though most of the functionalities remain the same, the functionalities are
carried out differently since the parameters monitored, the sensors used, etc., vary.
The process operation management of SAS has a completely different set of actions
to be performed including load control, fault identification, outage management, and
switching operations. Time synchronization is a functionality that the SAS must provide, for which a global positioning system (GPS) based clock is used to synchronize
the measurements and operations of all IEDs in the substation. The time synchronization system must ensure all the IEDs time stamp and perform all actions such as
measure, control, store and report the status and raise alarms for disturbances in real
time with resolution and accuracy of 1 ms. When all the schemes are employed in
all substations, the coordinated operation of all these units over a WAN gives way to
wide area protection systems (WAPS).

3.3.2.1 Communication Network for Substations

The SAS supports LAN and WAN for meeting the communication requirements of
all devices involved in satisfying the application demands. As all these communications happen within limited geographic locations, wired or even wireless communication technologies can be considered to establish the communication link among
the various devices. The most stringent performance requirements are in terms of
delay and packet delivery ratio (Table 3.11). Monitoring and data logging applications have some strong requirements in throughput also. There is the possibility of
ground potential rise in substations which generates high potential in the ground that
can damage conductive materials and so wired technologies often need expensive

TABLE 3.11

Communication Requirements in Substation (as per IEEE 1646)

Information Type	Internal to Substation
Protection messages	4 ms
Monitoring and control messages	16 ms
Operations and maintenance messages	1 s
Administration operations	1 s–1 min

and complex protection like multiple hard layers of sheath as in fiber optic cable. Wireless technologies have no such issue, yet the electromagnetic interference can deteriorate channel quality. WAN is provided to facilitate connectivity to neighboring substations for coordinated operations, to control centers or network operation centers for sending status updates and receiving commands governing operation of the substation and to energy markets. Delay and packet delivery ratio are the important metrics for these with strong requirement on throughput, also. Another very important use of WAN is the interconnection of all substations and control centers for achieving wide area situational awareness (WASA) by creating wide area measurement systems (WAMS) and also bettering control capabilities of the power system by realizing wide area control systems (WACS). These are not features of SAS and will be dealt with in detail in the upcoming sections.

3.3.3 Communication Requirements in Transmission

The transmission sector of a power system includes transmission substations and overhead transmission lines. A change in system state (aka, a disturbance) in one area easily gets cascaded up into a widespread problem with serious consequences. Information about events in neighboring areas can help utilities plan and optimize the economic operation of the grid. An important step toward transformation of the grid into a "smart grid" is wide area monitoring and situational awareness, as the inherent structure of the power system has so many interdependent and interconnected components distributed over a wide geographic area.

WASA refers to a set of technologies that equips grid operators with broad and real-time information about the functioning of the grid. Since all substations act as measurement points obtaining real-time operational parameters of the grid, by interconnecting substation SCADA and adding more IEDs, situational awareness of a wider area can be achieved. Typical SCADA systems provide voltage and current measurements every two to four seconds, requiring minimal latency as well as high reliability without being bandwidth intensive.

The choice of communication technology that materializes the interconnection of SCADA systems of multiple substations becomes more complicated in this context. For example, large communication delays during normal operations will reduce momentary situational awareness and control, affecting the performance of the systems only slightly. But large delays during a critical situation might stall the system and even collapse the entire grid. Also, the bandwidth requirements will grow as more nodes and services are made part of the system. Another major concern is the large coverage areas of transmission grid automation and the distance between substations and control centers as well as possible distance mismatch between individual stations or centers. Thus, the range and number of devices supported become important considerations in the selection of communication technology for a smart transmission grid. Some SCADA systems use combinations of licensed and unlicensed communication protocols; even satellite services may be used in remote or rural locations. Satellite services have universal coverage, but limited bandwidth capabilities and high latency. As with every other smart networked application, the SCADA systems are also moving toward IP enabled networks.

Synchrophasors – time synchronized values that give both magnitude and phase angle of a sinusoidal wave – are a key technological development that has redefined the entire style of wide area measurement and power system state estimation. Synchrophasors provide a long list of specific benefits to power systems including reliability and economic improvement, enhanced integration of intermittent and distributed energy resources, improved system modeling and planning, better contingency analysis and post event analysis of power disturbances, among others. Synchrophasor technology is utilized in a specialized device called phasor measurement unit (PMU) that gives accurate voltage and current phasor measurements with time stamps synchronized to a common GPS clock.

Phasor measurement units are devices that will be one of the keys to manifestation of a truly "smart grid." They sample power system signals at a very high rate, from widely distributed locations. The power system signals are then digitized in a PMU and fed into a processing unit for phasor computation. A phase locked oscillator along with a GPS reference clock provides microsecond level sampling accuracy to enable collection of large number of samples per cycle. The processing unit computes phasors and other parameters as per the IEEE C37.118.1-2011 specification and formatted as per the IEEE C37.118.2-2011 specification and these are sent to phasor data concentrators (PDC), then subsequently to higher level data concentrators (SuperPDC) located at control and monitoring centers or network operations centers (Figure 3.38). The PMU is not seen as technology to replace SCADA. Synchrophasor-based PMUs focus on widespread grid situational monitoring and awareness; whereas SCADA continues to be used for local automation with its monitoring and control capabilities. This helps in achieving the functionalities of a WAPS and WACS along with the WAMS capabilities fully realizing a wide area protection monitoring and control (WAMPAC) system.

Large scale deployment of PMUs is being carried out to provide a precise and comprehensive view of the power system, but difficulties arise in creating a uniform architecture for the system. These difficulties are mainly because of the intrinsic complexities of the existing power system like the presence of numerous utilities, dissimilar standards, overlapping territories, etc. As power transactions can take place anywhere in the network under free market regulation, a more centralized approach to network monitoring focused on vulnerability reduction of interconnected power

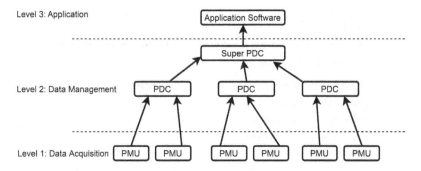

FIGURE 3.38 PMU-PDC network for WAMS.

systems is a must. Thus, it is evident that design of an open communication system that is reliable, robust, secure and based on interoperable standards, is one of the key issues in WAMPAC development and implementation. These ICT infrastructures should support ubiquitous interconnection with widespread and heterogeneous data sources, protection devices, decision making points and control devices; such infrastructure should be specified for varying levels of transport, security and dependability requirements in an open and interoperable framework.

3.3.3.1 Communication Network for Transmission Sector

A reliable and scalable communication backbone, that supports the various applications of emerging WAMPAC platform and large scale PMU deployment, needs a robust data transport WAN. The existing WANs for transmission sector are based on a hybrid mix of technologies including fiber optics, PLCC, GSM/GPRS, microwave and in some cases satellite communication. The communication requirements of WAMPAC varies depending on the nature of data being transmitted. Real-time monitoring and control requires low latency, typically in the range of 20–200 ms. For post-event, historical data, latency is not very vital. The typical operational data rate required is in the range of 600–1500 kbit/s. Over time, with the explosion of devices and the introduction of new applications, the cumulative bandwidth demands will increase. Reliability requirements are very stringent and backup power is required 24×7 (Table 3.12). The communication requirements of WAMPAC might drive the communication choices for other smart grid applications too. This is evident from the interest shown and investment made in low latency communications platforms by utilities. The implementation of WAMPAC is presently in the nascent stage and it is therefore not possible to pick a single technology as the best candidate as the choice needs more detailed field analysis on technical, economic and administrative options.

3.3.4 COMMUNICATION REQUIREMENTS IN DISTRIBUTION

Historically, the distribution side has had little "intelligence," as the job of the distribution grid was to distribute the electrical energy to different levels of consumers.

TABLE 3.12

Communication Requirements in Transmission Sector (as per IEEE 1646)

Information Types	External to Substation
Protection messages	8–12 ms
Monitoring and control messages	1 s
Operations and maintenance messages	10 s
Administration operations	1 s–10 min

The distribution sector has various levels of distribution substations, to step down the voltage, and overhead and underground distribution feeders. The distribution substations had SCADA systems for measurements and protection in the grid and substation automation. Control actions were taken in the grid based on the commands from control centers which are primarily based on measurements and data available from the transmission and central generation points. With the evolution of power systems and advent of technologies, an array of new applications and techniques have been developed to transform the distribution grid to a smart one. Probably, the most active sector in smart grid advancements in the first quarter of the twenty-first century is the distribution sector with applications like distribution automation (DA), distribution side management (DSM), distributed generation (DG), distributed energy resources (DER) that includes DG and energy storage, advanced metering infrastructure (AMI), demand response (DR), etc.

The DA includes remote monitoring and control of assets in the distribution network, like circuit breakers, switches, capacitors and transformers, operational control of grid, effective fault detection and power restoration, etc., through automated decision-making using distribution SCADA and distribution automation field equipment ranging from RTUs to IEDs. The additional functionalities provided by DA includes isolation of potential faults, raising alarm, automated feeder switching, improved fault detection, isolation and auto-restoration, improved capacitor bank and voltage regulator control, and communication of fault oscillographic data, all of which will help reduce the frequency and duration of customer outages.

The DSM is the process of modifying the demand for energy during peak hours by encouraging customers to use less or by shifting to off-peak times through various methods supported by financial incentives. DER extends beyond integrating RE-based generation and includes electric vehicles, combined heat and power, uninterruptible power supplies, utility scale energy storage, community energy storage, etc. AMI in distribution utilities facilitates to collect, measure and analyze electrical energy consumption data remotely for grid management, outage notification, billing and consumer information system via bidirectional communication; it can be leveraged to provide consumers with historical energy consumption data, energy use patterns, dynamic pricing information and incentives on reducing peak load. AMI also facilitates real-time net metering. The DR aims at reduction of electric

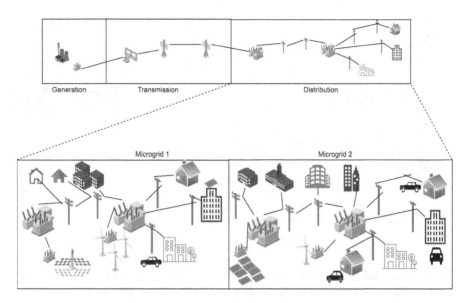

FIGURE 3.39 Smart distribution grid.

energy consumption during peak load periods and is typically operated by independent system operators and RTO for wholesale and by distribution utilities at retail level. These new energy technologies will modify the grid from the present unidirectional system to a grid with multi directional energy flow, from utility to home, home to utility, or even home to home, and bidirectional communication for every entity (Figure 3.39).

3.3.4.1 Communication Network for Distribution Sector

Various operational networks, like HAN – connecting all devices of a home, BAN – connecting all devices of a building, industrial area networks (IAN) – connecting all devices in an industry, FAN – network of field devices, NAN – network of HAN/BAN/IAN and any devices outside these networks and WAN – network of NAN and FAN, are created and maintained to meet the communication requirements of the distribution grid applications (Figure 3.40).

The latency requirements of DA applications vary widely – alarms and alert communications demand latency of less than one second, node to node communication demand sub-100 ms latency and many other applications accept latency up to two seconds. Bandwidth requirements are in the range of 9.6–100 kbit/s with stringent requirements of reliability and there is no need of backup power. The DER bandwidth requirements for individual distributed resources are around 9.6–56 kbit/s, while the latency requirements vary from 300 ms–15 s. In case of line faults and protection devices switching, the latency required is around 20 ms. The DER applications too need very high reliability and small backup power. Communication requirements of DR applications vary based on the level of system sophistication; basic level bandwidth requirement is quite low which increases to the order of 120 bytes per message

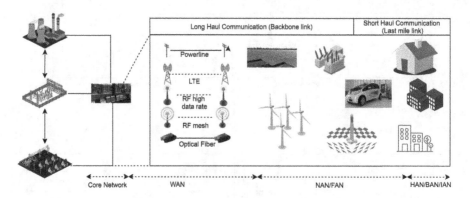

FIGURE 3.40 Smart distribution grid communication.

at medium level and it is in the range of 14–100 kbit/s per node/device at higher levels. The latency requirement also varies from as little as 500 ms to 2 s and even up to several minutes for different levels. Like other services, DR also has high reliability needs and does not need back up power (Table 3.13). Utilities use a variety of communications technologies like unlicensed wireless mesh, proprietary wireless, cellular as backhaul communications media followed by unlicensed 900 MHz spread spectrum and PLCC with unlicensed wireless to provide DA functionality. AMI normally uses PLCC with Wi-Fi, ZigBee, cooperative sensor networks and typically favors any incremental PAN supporting mesh network. The DR also uses PLCC with ZigBee or any other PAN. Some implementations of DR use broadband, next generation cellular techniques, Wi-Fi, or WiMAX.

TABLE 3.13
Communication Requirements in Distribution Sector

	Application	Data Size (bytes)	Latency	Reliability (%)
1	Meter reading	100 MB	15 s to 4 h	98–99.5
2	Pricing	100	1 min	98
4	Demand response	100	1 min	99.5
5	Distribution automation	100–1000	5 s	99.5
6	Outage and restoration management	25	20 s	98
7	Firmware updates	400–2000 K	2 min to 7 days	98
8	Program/Configuration update	25–50 K	5 min to 3 days	98
9	Customer information and messaging	50/200	15 s	99

3.3.5 COMMUNICATION REQUIREMENTS IN MICROGRIDS

A microgrid is smaller version of a legacy grid with constituent elements, like DER, distribution network, feeders, equipment for measurement, protection and control, control centers and loads, present within a fixed geographic boundary and act together as a single controllable entity. A typical microgrid connected to a legacy grid is shown in Figure 2.3. Chapter 2 has presented a detailed description of the design and operation of microgrids.

In grid-connected mode, the judicious control of microgrid assets will optimize local power generation, local consumption and also the demand charges for which the microgrid control systems should be able to communicate with all assets, market, operators, control centers, and other entities. The microgrid in islanded mode must have full WAMPAC capabilities, DER and SCADA systems for smart operation, providing all, and more, of smart distribution grid applications. The microgrid concept, in a way, changes the traditional organization of power system as generation, transmission, distribution and consumers into several types and sizes of microgrids, all interconnected to the main grid. A communication infrastructure to fulfill the large data exchange of these applications among end users and other stakeholders of the grid is hence a central subsystem of the smart microgrid. The role of a microgrid communication system depends on the entities that constitute the microgrid and the control system involved. Selection of communication technologies, configurations and protocols for microgrids depend primarily on the control objectives and costs of implementation and maintenance. Other design considerations include locations of nodes and communication characteristics like data traffic rates, degree of availability, and latency among other parameters. To achieve the high degree of availability of a communication system, its components should have redundancy. A typical structure of such a communication network has subsystems like HAN/BAN/IAN, FAN, NAN, and WAN (Figure 3.41).

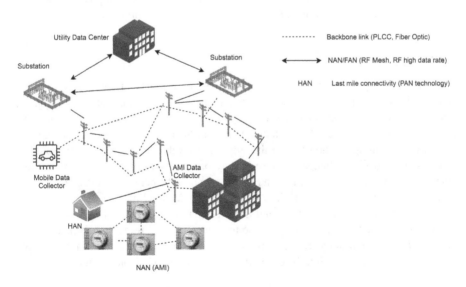

FIGURE 3.41 Microgrid communication network.

The best suited communication technology and its performance requirements are pretty much the same as presented in the previous sections as microgrid is a collection of all those.

3.4 STANDARDS FOR SMART GRID COMMUNICATION

Communication protocols provide a way to achieve data transfer between different nodes of a network usually developed to meet different types of needs. Communication protocols usually are defined in layers providing various functionalities essential for exchange of messages in a communication system. Some of the common layers across various protocols are discussed below.

Application layer protocols define message structures, services, information models and profiles, specifying types of data, their abstract formats and its translation into bits and vice versa. Transport layer protocols ensure synchronization among network elements in terms of message transfer rates and provide end-to-end delivery with network layer managing logical level addressing and mechanisms for navigating data packets through networks. Splitting long messages into smaller ones, correctly reassembling at the far end specifying style and timing for medium access with error detection and correction and physical device level addressing are implemented at link or medium access layers. The bottommost layer defines the bits to physical conversion and transmission styles of physical signal in the medium.

Smart grids encompass a large variety of connected devices to cater to the requirements of several applications. Several standards have been developed to ensure the smooth message passing between these devices thanks to the combined effort of several standard developing organizations like the IEEE, the ANSI, IEC, National Institute of Standards and Technology (NIST) and International Standards Organization (ISO). A comprehensive review of standards encompassing various aspects of smart grid communication is discussed in this section.

3.4.1 IEEE 1646-2004

The IEEE 1646 specifies the latency requirements for a variety of applications within and outside an electric substation for its automation (Table 3.14). The standard classifies substation communication into different categories because of the diversity of communication types and the variety of communication requirements of these types. For example, system protection application within the substation needs messages to be transmitted within 4 ms and file exchange allows a latency of up to 1 min. The standard defines *latency* as the time period that the data packets spend in the network between the applications running in the two end devices of the communication link. The *complete delay* includes the packet processing time and transmission time, and this should not exceed the required delay bound. As delay is due to processing and transmission phases, the standard defines communication capabilities like real-time support, message priority, data criticality, and system interfaces required to deliver information on time.

TABLE 3.14

Communication Requirements as per IEEE 1646

Information Type	Internal to Substation	External to Substation
Protection messages	4 ms (1/4 cycle of electrical wave)	8–12 ms
Monitoring and control messages	16 ms	1 s
Operations and maintenance messages	1 s	10 s
Text strings	2 s	10 s
Data files	10 s	30 s
Program files	1 min	10 min
Image files	10 s	1 min
Audio and video data streams	1 s	1 s

3.4.2 IEEE 1547.3-2007

The IEEE standard 1547, through its three parts – namely, the electric power system, the information exchange and the compliance test – defines the electric power system that interconnects distributed resources. The *power system* part specifies the requirements of different power conversion technologies, requirements of their interconnection and general guidelines on power quality, response to power system abnormalities and formation of subsystem islands. The *information exchange* part specifies the requirements on network aspects including interoperability, performance, extensibility and security issues for power system monitoring and control through data networks. The *conformance test* details the procedures to validate the compliance of any interconnection system to the standards and also describes a variety of tests to guarantee that an implemented system works as expected.

3.4.3 IEC 61850-90-5

IEC 61850 specifies a standardized framework for substation automation and integration, developed by the IEC Technical Committee 57 – Working Group 10. It specifies the communication requirements, the functional characteristics, the structure of data in devices, the way applications interact and control the associated devices and methods to test conformity to the standard. The standard is described in ten separate parts and IEC 61850-90-5 is the part which deals primarily with transmission of synchrophasor measurement data. IEC 61850 uses the common information model of real devices in terms of logical nodes that can define various functionalities in the implementation of real devices (Figure 3.42).

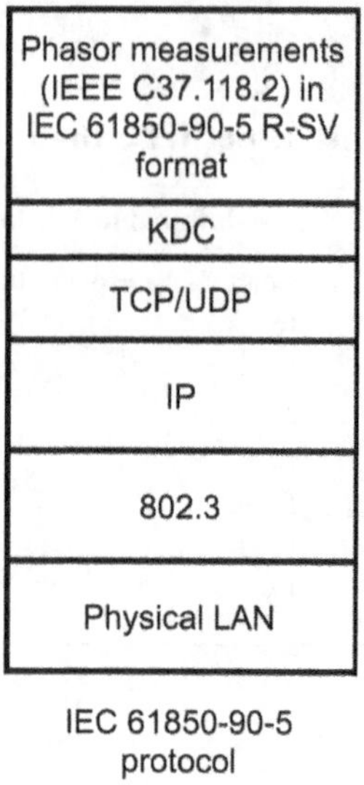

Phasor measurements (IEEE C37.118.2) in IEC 61850-90-5 R-SV format
KDC
TCP/UDP
IP
802.3
Physical LAN

IEC 61850-90-5
protocol

FIGURE 3.42   IEC 61850-90-5 protocol stack.

Information modeling provides standardized syntax, semantics and hierarchical structures for the data exchange among multiple devices and systems. Groups of data objects that serve specific functions are called logical nodes and several logical nodes combine to form a logical device. IEDs are physical devices having one or more logical devices and PMU is a type of IED with relevant logical nodes for synchrophasor measurements. New logical nodes and data objects are added for modeling PMU to incorporate requirements like rate of change of frequency, sampling rates, class of PMU, etc., to the existing IEC 61850 compliant merging units whose model contains the logical nodes for measurement and operation related information. Also many logical nodes addressing common issues of all logical devices, providing physical information, providing the measured phasor angle, etc., are added in the PMU information model.

Service modeling takes care of handling a particular IED's different message exchanges or services. The PMU message of synchrophasor measurements is mapped onto a routable UDP/TCP service having multicast capabilities by using routable sample values and generic object oriented substation events obtained after adding network and transport layers to the conventional versions. Control blocks for managing the data transfer are created by adding new functional constraints to the existing control blocks of IEC 61850-9-2 and 8-1 standards. The new functional constraints add provisions to incorporate the IP protocol in conventional control blocks by adding new attributes containing priority, IP address, IP type of service, etc.

The service exchanges between PMU and PDC following IEC 61850-90-5 model happens as shown in Figure 3.43. The PDC sends a manufacturing message specification – request type message, similar to the command for configuration frame in IEEE C37.118.2, to obtain the information model of the PMU. PMU responds with a manufacturing message specification – response type message with the data of all the logical nodes. The PDC gives a start command via the control blocks and PMU

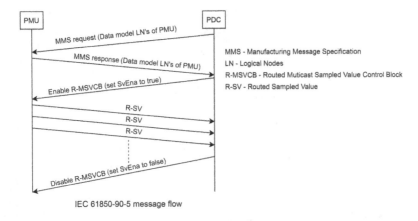

IEC 61850-90-5 message flow

FIGURE 3.43   IEC 61850-90-5 message flow.

continuously transmits messages with sample values to PDC till the turn off transmission command is send by the PDC again via the control blocks.

3.4.4   IEEE C37.118.2

The IEEE C37.118.2-2011 defines real-time synchrophasor data transfer between the PMUs and the PDC, and it is evolved from IEEE 1344-1995 and the IEEE C37.118-2005. Synchrophasor communication network has several PMUs distributed over a large geographical area. The PMUs are capable to continuously measure the voltage and current values, time synchronize using the reference GPS clocks, compute phasors and send these time-tagged phasors to the PDC, SuperPDC and control center for continuous and real-time monitoring and control (Figure 3.44).

A well-defined message format and sequence of messages are required for communication among devices developed by the different vendors. The synchrophasor standard defines four types of messages – namely, data, configuration, header and command. Command messages are sent from PDC to PMU for giving various commands for configuration and control. Data messages are sent from PMU to PDC which contain information about the synchrophasor measurements and various other status parameters and measurements. Configuration messages are also sent from PMU to PDC and these have information pertaining to the data types, data calibration and other metadata. Header messages are again from PMU to PDC and these contain human readable information provided by the user. The data message format is shown in Figure 3.45.

The IEEE C37.118.2 data message starts with a sync field that specifies the frame type and version number and the next field gives the frame size in bytes. The next three fields give the source ID, second of century, and fractional second of century information. The data field has phasors, frequency, deviation in frequency, analog and digital values and is followed by a checksum field. The sequence of message exchanges for data transfer between PMU and PDC as per IEEE C37.118 is shown

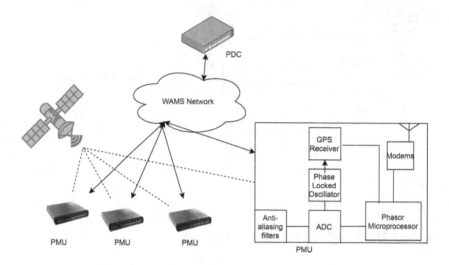

FIGURE 3.44   PMU – PDC communication link based on IEEE C37.118.2.

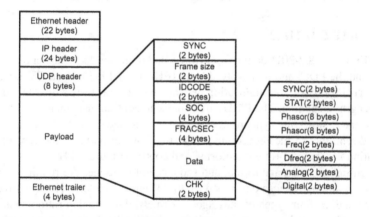

FIGURE 3.45   Data message frame format in IEEE C37.118.2.

in Figure 3.46. The first is a command message from PDC to PMU requesting for its configuration to which PMU responds by sending the specific configuration frame, following which the PDC again sends a command message requesting PMU to turn on data transmission. The PMU flushes data packets to PDC repeatedly until PDC sends turn off transmission command.

3.4.5   IEEE 1815-2012

Distributed network protocol 3 (DNP3), developed by Westronic Inc. in 1992, standardized as IEEE 1815 in 2010 and superseded by IEEE 1815-2012 standard, specifies standards for communication between the control stations and the

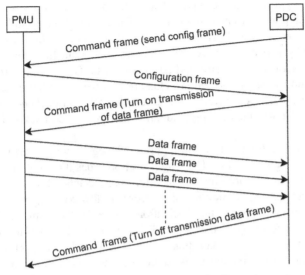

IEEE C37.118.2 message flow

FIGURE 3.46 IEEE C37.118.2 message flow.

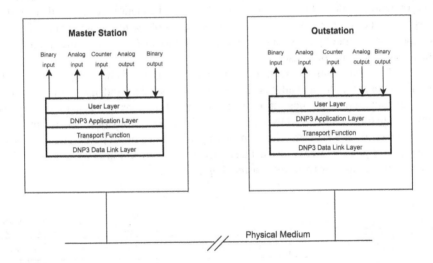

FIGURE 3.47 Master – Outstation interaction model – DNP3.

control equipment in electric power systems. It has a crucial part in providing communication procedures in SCADA system, between SCADA master and RTU and IED. The master and outstation interaction model used by the DNP3 is shown in the Figure 3.47.

User layer of master control station initiates the data transfer making the application layer to send an initiation request to the outstation, carrying the information regarding the type of required data. The transport layer, upon reception of the request from application layer, segments it. The data link layer adds information relating to the address and error detection and the data packets are transmitted through the physical medium. At the outstation, the data link layer receives the packets and performs both error and address checks. On passing error detection, the address and the error correction fields are removed and the remaining packet is handed to the transport layer which assembles all data packets belonging to a single request. The application layer identifies the requested data and gets it from the user layer.

DNP3's data link frame, shown in Figure 3.48, begins with a start field that indicates the beginning of a frame. The *length field* provides the length in octets of the complete DNP3 frame whose maximum value is 292 octets. The *control field* specifies frames direction, transaction initiator, error and flow control, and function code. The destination field presents the destination device for the frame, and the source field has the source of the frame. The CRC field is used to provide integrity for the other eight octets of the data link frame. The pseudo-transport segment consists of 3 fields: FIN – indicates last segment, FIR – indicates first segment and Sequence – incremental counter to ensure segments are not duplicated or missing and are in order. DNP3 application fragment begins with an application control field that has FIR, FIN, *confirmation* flag-used by receiver to indicate a confirmation message, *unsolicited* flag to show whether response is unsolicited or not and sequence. The Function Code field uses one of the 34 defined function codes to identify the purpose of the fragment in requests and responses. Following the application header are the DNP3 data objects which are used to send to, and request information from, the master and slave databases.

DNP3 provides multiplexing, data fragmentation, error checking, link control, prioritization, and physical addressing services for user data and also defines a transport function and an application layer that defines functions and generic data types appropriate for normal SCADA applications. The DNP3 protocol supports time synchronization with a RTU and has time stamped variants of all point data objects so that it is still possible to reconstruct a sequence of events to know what happened in between the data polls from the master (Figure 3.49).

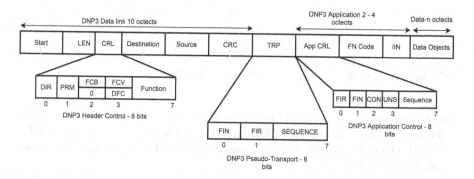

FIGURE 3.48 DNP3 data frame format.

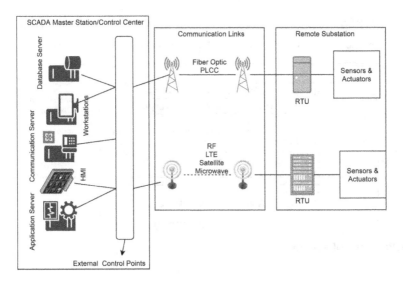

FIGURE 3.49   A sample DNP3 system architecture.

3.4.6   DLMS/COSEM – IEC 62056

The IEC 62056, successor of IEC 61107, is a widely used protocol for exchange of metering data defined by IEC based on the device language message specification (DLMS) that defines the communication profile and the data objects, and the procedures for information exchange between the devices are specified by the companion specification for energy metering (COSEM) that defines the transport and application layers of the protocol. The DLMS/COSEM specifies the communication standard across a variety of communication media and has various subclasses as defined in four technical reports – namely, Green Book, Yellow Book, Blue Book and White Book. The Blue Book details the COSEM meter object model and the object identification system (OBIS), the Green Book presents the architecture and protocols, the Yellow Book is on conformance testing, and the White Book contains glossary of terms. The DLMS/COSEM protocol can be used for a variety of different metering systems including electricity, gas, water, etc (Figure 3.50).

Object modeling is a potent tool in formally representing data with each aspect of the data modeled with an attribute. Objects can have many attributes and methods to perform the operations. Modeling of simple use cases such as register reading and more complex cases as tariff and billing schemes or even load management can be achieved by using object combinations. The COSEM object model describes the semantics of the language and is named by OBIS. There exists 281,474 billion OBIS codes specifying electricity, gas, water, heat cost allocators and thermal energy metering, as well as abstract data whose hierarchical structure allows classifying the characteristics of data. Around 4,398 billion OBIS codes are reserved for standardization purposes and the rest is used for manufacturer, country and consortia specific purposes.

DLMS/COSEM uses a client-server model of data exchange with the meters acting as the servers and the control centers or data concentrators acting as the clients.

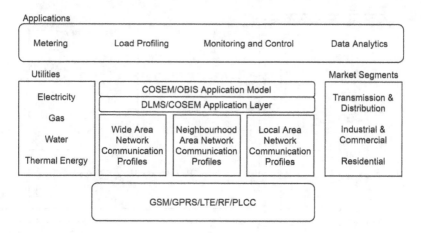

FIGURE 3.50   DLMS overview.

The DLMS/COSEM application layer provides services to interconnect clients and the servers and to access the data held by the COSEM objects. The application layer creates messages, referred as application protocol data units, manages cryptographic protection including applying, checking and removing as needed and handles transfer of long messages in blocks. The message transfer is supported over many types of communication media and several mechanisms are available for optimizing the traffic according to the media characteristics.

The DLMS/COSEM application layer has services that can be used to implement the communication needs of the metering device which acts like a data server and the data concentrator or the utility center acts like a data client. The application layer provides services such as *connection management* involving connection establishment, connection abortion and connection release, and *data communication* involving data reading, data writing and method calling, for the application processes and link layer. The LLC sublayer checks data flow direction allowing compatibility between connectionless and connection-oriented layers. The MAC sublayer specifies services and protocols for medium access and uses high level data link control (HDLC) frame format type 3. The protocol supports any physical layer implementation. Client application processes, through a client application layer service, requests the server application processes for data collection or any other action. The server application process decides the object to be invoked. The data or the result is sent back to the client application process using a server application layer service.

The LLC data format has four fields. The destination LLC service access point is usually fixed and is used for broadcasting. Source LLC service access point is also fixed and the last bit is used as a command or response identifier. LLC_Quality is reserved for future use and the information field carries the LLC service data units. The HDLC format used in MAC sublayer has nine fields and the opening and ending flag fields always have the value 0 × 7e. The frame format field has four bits for format type, one bit for frame segmentation and 11 bits for frame length. The destination and source addresses are MAC addresses of devices involved. The control

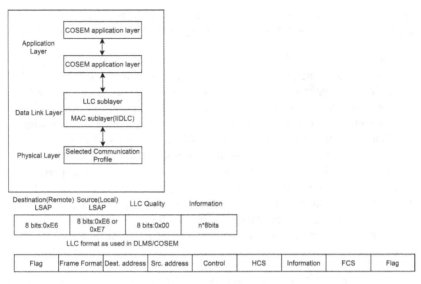

FIGURE 3.51 DLMS/COSEM protocol stack and data format.

field indicates the type of commands and responses. Header check sequence (HCS) and frame check sequence (FCS) are 2-byte fields. The information field carries the payload and in case of data frames, it has application protocol data unit (Figure 3.51).

3.4.7 IEEE 2030

A major step in integrating all the extremely important experiences in providing alternate approaches and best practices for realizing interoperability in smart grid is the "IEEE 2030 – Guide for Smart Grid Interoperability of Energy Technology and Information Technology Operation with the Electric Power System (EPS), End-Use Applications, and Loads." It is the first all-inclusive IEEE standard on smart grid interoperability specifying a roadmap to establish the framework in developing a body of standards based on multiple and overlapping technical disciplines in power system applications, information exchange services and control through communications.

IEEE 2030 sets up the smart grid interoperability reference model (SGIRM) and presents a knowledge base addressing nomenclature, attributes, functionalities, performance, assessment criteria and usage of engineering principles for smart grid interoperability. Smart grid interoperability based on a *system of systems* approach is the foundation on which IEEE 2030 establishes SGIRM as a design that essentially allows extensibility, scalability, and upgradeability. A conceptual representation of the smart grid architecture with three integrated architectural perspectives (IAP) in power systems, communications technology, information technology and characteristics of the data is defined in the IEEE 2030 SGIRM with details on design tables and classification of data flow characteristics essential for achieving interoperability. The aim of each architectural perspective is to address interoperability among the smart grid entities. Each perspective contains unique characteristics addressed from its respective technology function. A summary of the three IAPs are as follows:

- *Power systems IAP (PS-IAP)*: The focus of this perspective is the generation, transportation, delivery and consumption of electric energy including equipment, applications and operational ideas. This perspective presents seven domains common to all perspectives: Bulk generation, transmission, distribution, service providers, markets, control/operations, and customers.
- *Communication technology IAP (CT-IAP)*: The focus of this perspective is communication facility among systems, devices and applications in the context of the smart grid facilitating data exchange. The perspective includes communication networks, media, performance, and protocols.
- *Information technology IAP (IT-IAP)*: The focus of this perspective is the control of processes and data management flow. The perspective includes technologies that store, process, manage, and control the secure information data flow.

Entities are placed inside one of the domains and are interconnected to each other using one or more of the specified interfaces. Each perspective has entities that are more closely related to its respective technology and each entity can map to any appropriate entity or entities in any perspective through interfaces which are logical connections that support data flows. The reference model given is functional, expandable and not expected to be insistent or constraining. It is natural to need to maintain interoperability as smart grid technologies and architectures evolve and the reference model is flexible to accommodate smart grid evolutions well into the future (Figure 3.52).

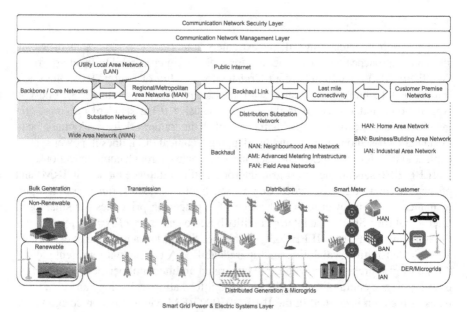

FIGURE 3.52 End-to-end communication model as per IEEE 2030.

3.5 COMMUNICATION INFRASTRUCTURE FOR SMART MICROGRIDS

A communication infrastructure interconnecting, almost, each and every entity, coming under all stakeholders, facilitating bi-directional information flow is essential for realizing a smart microgrid. This is the backbone system that has to meet the data transfer requirements of all various smart microgrid applications with extremely varying demands and does so with the capabilities of heterogeneous and interoperable communication architecture.

3.5.1 NEED FOR COMMUNICATION ARCHITECTURE

Communication architecture is a framework that defines the role and functionality of each participating device in terms of message transfer and network organization to realize the data communication. There exists a variety of communication architectures and the choice or deployment depends on the application to be catered to (Table 3.15).

There is a growing consensus amongst smart grid stakeholders on features such as scalability, interoperability, availability, wide usage and reliability, latency, bandwidth, throughput and packet delivery ratio as the deciding factors on choice of communication technology for creating the network. Specifically, the smart grid activities include measurements of energy consumption and operational details, controlling energy provision, and monitoring and control of generation and distribution according to the power flow requirements. The communication relationships are not always peer-to-peer, but more of a multi-client multi-server nature. As seen from the previous two sections, each and every application has communication path defined by the respective standard and the hierarchical network levels are used to realize the message passing. As so many applications with widely varying operational styles and requirements exist in the same geographical area, the optimal communication network setup and message exchange is not a straightforward task; it needs a reliable and scalable communication architecture for progressive realization of the smart grid. Selection of the best-fit communication technology to implement the architecture is subject to many technical, legal and strategic restrictions.

TABLE 3.15
Features of Communication Network

Feature	Definition
Scalability	Dynamic handling of the topology based on number of nodes.
Ubiquity	Access grant to authorized users from any network point.
Interoperability	Support to all types of networks or devices.
Integrity	Guarantees round the clock operation.
Standardization	Well defined interconnection for different network elements.
Upgradeability	Upgradation of software, algorithms, and system configuration.

3.5.2 HIERARCHICAL COMMUNICATION ARCHITECTURE IN SMART GRID

The smart grid communication network is composed of various smaller communication networks intended for realizing data exchange for a variety of applications that exist at various sectors of the grid and serve the stakeholders concerned (Figure 3.53).

1. *HAN/BAN/IAN*: HAN is a network of all smart devices in a home for smart operation of appliances. The smart devices include those which participate in DR or that can be remotely turned ON/OFF via commands, etc. Smart meter is by default a member of HAN network. BAN is a version of HAN that extends to a building and may have more than one smart meter each belonging to a different consumer. BAN also has smart devices other than smart meters which operate based on commands from respective home controller, building level controller or distribution utility. IAN is similar to a BAN in the context of an industry and may or may not have more than one smart meter and has a lot of smart devices. HAN/BAN/IAN forms the network of data collection and actuation points among the various consumer classes and is a critical element of the smart grid communication network. Both data collection and control are achieved through these networks which hierarchically take the bottom position; these networks are usually

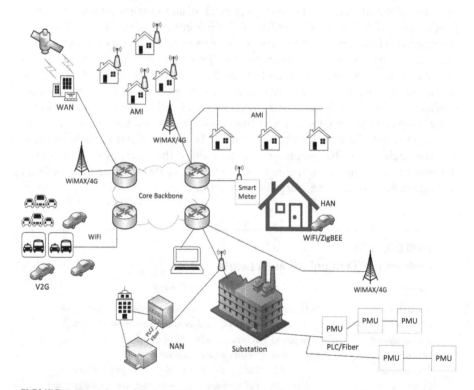

FIGURE 3.53 Hierarchical communication network for smart microgrid.

fulfilling the *last mile connectivity* and these realize smart grid applications like AMI, DR, DSM, etc. PLCC is the widely used technology in many utilities and equally or even more preferred are the wireless counterparts in low data rate RF and other PAN, and wireless LAN standards.

2. *FAN*: FAN is the network of field devices distributed throughout the traditional transmission and distribution sectors, at substations and locations of installation of outdoor equipment like distribution trans-formers, for measurement, protection and control operations. This is a network in which end devices play a vital role in operation of the grid and is part of WAMPAC (sometimes in standalone microgrids and even in DER) realizing its capabilities by further getting integrated to NAN and WAN or to WAN directly. The network nodes act as source nodes in realizing WAMS sharing measurement data and WAPS and WACS sharing status data and also act as end actuators in realizing WAPS and WACS acting based on intelligence and commands from network operations and control centers. This network sometimes operates as a heterogeneous one with low bandwidth and low latency interconnecting the field devices to the local master; while highly reliable technologies like PLCC and fiber optics act as the back-up connecting FAN to NAN or WAN.

3. *NAN*: NAN is a network interconnecting smart meters and data concen-trators and is often considered as the last mile communication link of the smart grid communication network. The number of smart meters under a concentrator varies and depends on the communication technology used and network topology employed. HAN/BAN/IAN are connected to NAN to share the data generated in one direction and to receive the commands and configurations in the other providing a gateway facility for the utility's WAN to access the consumer's devices. Typical NAN operation is within a communication range of 10 km supporting data rates in the range of 10–1000 kbit/s and employs mesh networks based on IEEE 802.11s, IEEE 802.15.4g and other PAN and wireless LAN technologies offering self-configuration, high speed, and reliability.

4. *WAN*: WAN implements a bi-directional backbone communication link across various sectors and for all the smart grid applications. Interconnections are provided between utility control centers, markets, service provider data centers, NAN and FAN for exchange of voluminous data in one direc-tion and control and configuration commands in the other through high-bandwidth media. The WAN covers a range of 10–100 km and fiber optics forms the first choice for this network, being robust and having very high transmission capacity, although costly to deploy. Alternative technologies include WiMAX, broadband over PLCC, etc.

5. *SCADA network*: This involves networking among various devices in a substation or generation plant or control center for automation of the entity concerned. This has limited bandwidth requirements, but quite stringent latency and reliability demands. Typical choice of technology includes Ethernet, fiber optic, and wireless LAN.

3.5.3 Communication Messages in Smart Grid

The messages in smart grid communication network can be classified as monitoring/
status update, command/configuration and administrative at various levels of operation of the grid.

1. *Monitoring/Status update messages*: The very core feature that makes
   smart grid a reality is the capability of all smart grid applications to make
   intelligent decisions in real time. The prime necessity to achieve this is
   data. Smart grid at its various levels of operation generates humongous
   volumes of data which the resource-constrained nodes – distributed for
   data collection and control – cannot process and use for real-time decision making. There are some applications in smart grid, especially related
   to protection, where sensing, processing and action are done locally but
   local processing is not true for the major share of applications. Even in the
   case of protection, configuration and status, update messages need to flow
   in both directions. One major contributor to measurement data is a smart
   meter which is present at every point of consumption and measures the consumption data along with power quality data. These data are passed to the
   distribution utility centers, operation centers and data centers via several
   hierarchical layers of data concentrators depending on the area covered and
   metering standards in force. The number of intermediate devices is also
   dependent on network topology and choice of communication technology.
   The smart meter network typically uses HAN and NAN for achieving its
   connectivity and sometimes uses even WAN. Another major contributor to
   data monitoring is IED distributed in the grid for real-time measurement
   like PMU. These devices measure electrical signal parameters, do processing in terms of phasor computation, time tagging, etc., and include status
   messages of various components (like circuit breakers and relays) along
   with the data message. The PMU data flows via several hierarchical layers of PDC before it reaches the utility centers. The number and layers of
   PDC will vary based on the involved WAMS/WAPS standards. The data
   flow between these devices, utility operational centers, data centers and
   markets happen through FAN and WAN. The various SCADA networks
   deployed for automation at substations and power plants also do measurements. The data originating from the end devices pass through hierarchical
   layers of master nodes and gateway before it is passed on to utility centers.
   The number and levels of master nodes change based on the level and type
   of automation and the plant and associated standards. Even though the processing and actions needed for automation are done locally, the related data
   for real-time operation, control, analysis and storage are sent to the utility
   centers using LAN and WAN.
2. *Command/Configuration messages*: Upon receiving the information
   through the messages, the next step is initiation and execution of control
   actions within the time limits specified in the respective standards. The control signals are generated at different locations or levels of operational and

control centers, and sometimes even from local field level masters. Another type of data that flows from the utility control centers to the individual devices is the configuration for operational specifications of all measurement, protection and actuation devices in the grid. Sometimes the configuration messages flow in the opposite direction (i.e., from device to the control centers specifying the current configuration of the device). The hierarchical layers in the path and the delivery time limits for each message are all specified in the respective standards. The major command and configuration message exchange happens between IEDs (like PMU) and utility centers which are critical for realizing operational capabilities of WACS and configuration of WAMPAC and is achieved using WAN and FAN. Another share of command and configuration data exchange is between the smart meters and utility centers realizing DA, DSM and DR capabilities; also, between SCADA systems of distributed generation and storage systems as well as utility centers to realize DER capabilities and achieved using WAN, NAN and LAN. Similar message exchange exists between SCADA systems of substations and power generation plants for certain control and configuration actions, and sometimes based on inputs from control and operation centers and this is achieved using WAN and LAN.

3. *Administrative messages*: There is a need for message exchange among entities such as customers, DER, substations, generating stations, control centers, markets and service providers which are essential for the necessary governance in executing the right operational strategy for the power grid. This type of message exchange does not have a stringent requirement on latency, but sometimes the exchange demands large bandwidth availability for huge amount of data transfer for analysis and post-mortem. Such messages exchanged between consumers and utility help in developing the consumer information system, sharing consent on DR, information on multiple tariff schemes, operational limits for any DG present, billing information, incentives, etc. AMI administrative message service also includes request for measurement of energy data, service restoration and disconnection, firmware update, etc. The control centers, substations, generating stations and markets exchange administrative messages for energy trading, analysis of operational performance, and service management.

3.5.4 Communication Data Path for Smart Grid Applications

Communication data paths of some popular smart grid applications are presented below (Figure 3.54).

1. *AMI Communication Path*: AMI has transformed over the years replacing the automatic meter reading (AMR), which used one-way communication for monthly or bi-monthly meter reading, data transfer and billing; AMI implements two way flow of data using the hierarchically connected networks such as HAN, NAN and WAN. Configuration commands and control

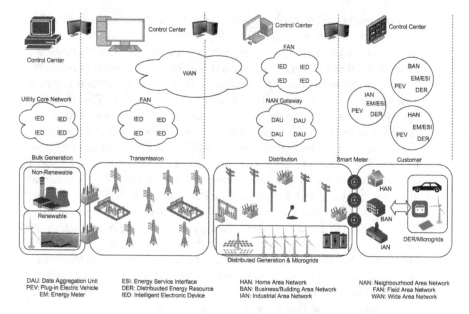

FIGURE 3.54 Communication data path for smart grid applications.

messages flow from the utility centers to the smart meters at the consumer premises, and measurement and status update or event messages are sent from the consumers to the utility. AMI communication path can be mapped to the WAN-NAN-HAN inter network in many ways based on the capability of the technology of choice and the network topologies supported by the technology. PLCC can form a network with tree topology with immediate hops to the data concentrator, while a radio technology can create mesh topology with multiple hops to the data concentrators. A heterogeneous network with PLCC between utility center and data concentrator, and wireless technologies such as IEEE 802.15.4g between data concentrator and smart meters will need data concentrator to be a multi-port device supporting both the technologies. The periodic, on-demand or event driven data from the numerous smart meters can be coalesced at the energy service interface (ESI) to serve a lot of applications including meter data management system, outage management system, asset management system, distribution management system and customer information system which are implemented at the utility center servers via enterprise software.

2. *DR Communication Path*: The DR schemes are developed by utilities to alter the energy consumption profile of all consumers to bring down peak demand in response to incentives or dynamic pricing. The DR does not have stringent communication demands in latency, availability, reliability and bandwidth for commissioning and enrollment processing of devices in the ESI. But, there are critical communication demands on performance metrics for direct load control messages from the utility. The DR can use IEEE

802.15.4, IEEE 802.11 and IEEE 802.3 standards for LAN and wireless data transfer over cellular network, PLCC or a heterogeneous system with PAN, wireless LAN and other high bandwidth backbone technologies for NAN and WAN. It is usually implemented over the AMI network as the requirements match.

3. *Electric Vehicle (EV) Communication Path*: An EV, plugged in to the grid at customer premises can be treated as smart load or source in the HAN. The EV interacts with the grid in two modes: grid to vehicle (G2V) for charging and vehicle to grid (V2G) to act as a power source via the electric vehicle supply equipment (EVSE). The communication between vehicle battery and EVSE happens over controller area network (CAN) connection or PLCC following GB/T, CHAdeMO, IEC 15118 or other proprietary protocols. Further, the EVSE needs to talk to the utility server following open charge point protocol (OCPP). The connection between the ESI and EVSE is established through technologies such as IEEE 802.3, and IEEE 802.15.4 for LAN and can use any technology for WAN. This is also implemented over the AMI network as the requirements can easily be satisfied.

4. *DER Communication Path*: The DER represents variable renewable energy sources and storage systems distributed across the energy network. The intermittent nature of generation by DER will affect the system balance and causes service degradation, and therefore needs IED for monitoring and controlling including islanding, analyzing, protecting, starting and stopping the DER. In the case of DER at consumer premises a net energy meter with ESI facilitates monitoring and control over the AMI network and provides the communication path. In the case of DER at distributed power plants with LAN interconnecting SCADA system and IED, having gateway nodes to interconnect to WAN, implements the communication path.

5. *WAMPAC Communication Path*: The WAMPAC includes time-synchronized display of the electrical state and performance analysis, protective actions and providing the appropriate response system events to enable optimal system operation. PMUs in large numbers are deployed in the transmission and distribution domain for WAMPAC which measure line current, bus voltages and frequency to estimate the health and integrity of the grid. The huge amount of data generated by PMU is sent to control centers via hierarchical layers of PDC and hence, would require very high bandwidth link for PMU-PDC-Super PDC-control center connection realizing the WAN. The typical choices are fiber optics, PLCC and WiMAX.

6. *Distribution Grid Management (DGM) Communication Path*: The DGM covers DA and almost all other smart distribution grid applications and uses HAN, NAN, FAN and WAN to achieve its functionalities. The DGM is a smart microgrid system without islanding management and its various services use all the communication paths that are discussed so far. The ICT applications in DSM – namely, DR and demand dispatch – are explained elaborately in Chapter 4.

3.5.5 Communication Network Requirements

Communication networks to be suitable for smart grid must meet distinct QoS requirements demanded by the applications. The major requirements are:

1. High *reliability* and *availability* are common requirements for almost all smart grid applications, including some of the administrative aspects. The requirement is that all nodes must always be reachable, thereby ensuring that there are no link failures in the network. This is not an issue in a wired network, but reliability and availability may be challenging in a wireless or powerline system. Just as a high level of interference may pose a problem in wireless networks, the switching actions by SCADA and control centers (or even manual changes) that impact network topology is a problem in PLCC. Robust and dynamic network configuration and redundancy can be a solution to this problem.

2. Automatic management of *redundancies* aims at ensuring availability by activation of hot standbys at node and link levels through dynamic network management to facilitate communication paths for time critical applications during node faults, link failures and even topology changes.

3. A requirement of WAN technologies is high coverage distances or *range*, as the nodes to be interconnected by the network are distributed over a wide geographical area. Technologies that satisfy this requirement include telecommunication systems, PLCC, fiber optic, etc.

4. NAN and WAN technologies, and sometimes FAN too, require the ability to support large *number of nodes*. The best suited are IPv6-based technologies if unique addressing is intended for all nodes of the network, else even IPv4 can satisfy typical requirements of sector wise networks.

5. Communication *delay* and system responsiveness is a major requirement of all message types in smart grid because of the dynamic and critical nature of the system. The system responsiveness demands are different for different data classes such as metering, status update, control and administrative which range from a few milliseconds to minutes and even hours.

6. *Bandwidth* is another critical requirement for technologies used to create the backbone network like NAN and WAN. HAN and FAN might not have pressing demands on bandwidth, but it is important for all monitoring applications and sometimes for configuration and administrative applications.

7. Power system being considered as a critical infrastructure makes associated data also critical and hence communication *security* is yet another important requirement. Lack of security in communication can lead to grid collapse besides privacy, integrity and authenticity related issues. Secure communication is therefore important.

8. *Ease of deployment and maintenance* is a common requirement of communication technologies used for any application and the case in smart grid is not different. For any distributed communication system, methods to facilitate initial installation and maintenance of the infrastructure must be foreseen and provisions must be provided. Error localization, analysis, easy and remote update of firmware and remote configuration are some of the essential features.

3.5.6 COMMUNICATION NETWORK VALIDATION

Traditionally, the performance analysis of power system and ICT is done separately. In the context of the smart grid, ICT and power system are seamlessly integrated and the impact of ICT in deciding the performance demands simultaneous analysis of both in a comprehensive system level. Both domains employ simulation tools traditionally for performance analysis and the simulation complexity is in the increase with the extension of the network in both cases. The detailed analysis in depth as seen in a real-time hardware in the loop (HIL) simulation is not possible in separate and individual software simulations. The performance gets considerably affected as one simulation tool gets coupled with the simulation tool of the other domain owing to necessary overhead involved in data exchange and format conversion. A simulation involves development and implementation of all component models in a system whose performance analysis is to be carried out or proof of concept has to be validated. If models of the two systems are developed together but are executed separately, it is called a distributed or parallel simulation approach. Execution of multiple models in the same software environment with different modeling styles is called hybrid simulation. The overview of various approaches for performance validation used in the energy domain with ICT like (1) multi domain simulation, (2) cooperative simulation, and (3) HIL simulation, are discussed below.

1. *Multi Domain Simulation*: Simulation tools exist in all domains for design and proof of concept validation. For concept analysis, analytical or numerical tools are suitable, provided there is clear separation of the solutions for analysis of the multiple domains involved, otherwise the simulation results will not be useful for validation of the system. Modeling the domains involved and focusing on the determinism of simulation outcome is a way of achieving performance validation of multi domain systems via simulation. There exists multi domain simulators that adopt object oriented and declarative approach, mathematical approaches, etc., and are built on generalized principles of interaction. Typically a multi domain simulator would want the system representation under consideration to be altered to suit the simulator architecture.

2. *Cooperative Simulation (Co-simulation)*: Co-simulation is defined as the synchronized implementation of two or more models that vary in their underlying modeling paradigm as well as in the software system that solves or executes the model. The models in a co-simulation are developed and implemented individually implying the interaction of hardware and software components. The major advantage of co-simulation is the separation achieved in modeling and simulation. Most co-simulation approaches employ manual coupling of two software executing models of individual systems normally via a middleware software called co-simulation framework that is responsible for information exchange and temporal synchronization between the two models. A co-simulation framework offers interfaces for connecting a model using which the model can achieve data exchange with all other connected models in an indirect fashion. The framework

provides a synchronization algorithm facilitating a common time frame for data interchange avoiding the need of individual linking between simulation tools and reducing the chances of coupling errors.

3. *HIL Simulation*: HIL will have real-time simulation where the time step for achieving a solution and accomplishing I/O activities must match the real-world time requirements. These will commonly be coupled with a hardware unit which is concurrently being tested. HIL allows thorough analysis of the hardware while the simulation allows testing of a wide range of operating conditions, decreasing the risk of performing tests on an actual network as well as the cost and time required.

3.6 CASE STUDY

This case study presents the communication network designed, developed and implemented for smart operational facilities in a laboratory scale hardware simulator of microgrid developed at Amrita School of Engineering, Coimbatore, India. The microgrid envisaged through this simulator delivers electricity over a geographical area of nearly 25 km² with a peak power demand of 15 MW.

The microgrid has distributed power plants such as: (i) 15 MW micro hydroelectric plant, (ii) 6 MW wind farm, and (iii) 625 kWp solar PV farm. The microgrid is synchronously tied to the main power grid which supplies the deficit of local generation in meeting the local demand and absorbs the excess power locally generated; when required it can be operated in autonomous mode too. Figure 3.55 depicts the single line diagram of the 5 bus microgrid. The microgrid under consideration operates at 11 kV, which is quite high for a laboratory scale emulator. Therefore, the microgrid parameters were scaled down to 400 V with 1 kVA as base power to form the laboratory model. Table 3.16 shows a comparison of the actual microgrid components with the scaled down version (Figure 3.56).

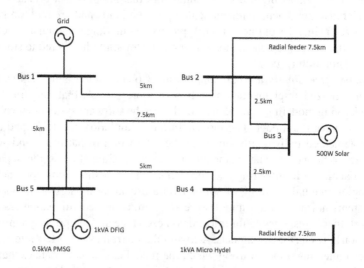

FIGURE 3.55 Single line diagram – 5 bus microgrid laboratory model.

TABLE 3.16
Microgrid System Description

S.No.	Actual System	Scaled Down System
1	15 MW micro hydel power plant (MHPP)	1 kVA synchronous generator driven by DC motor
2	6 MW wind power plant (WPP)	0.5 kW Wind turbine generator (WTG) comprised of (i) 1 kW wind turbine emulator, and (ii) 0.5 kW permanent magnet synchronous generator (PMSG)
3	625 kWp solar photo voltaic (SPV) plant	500 Wp SPV panel
4	11 kV distribution line	Lumped line sections, each unit representing 2.5 km of distribution line, suitably interconnected

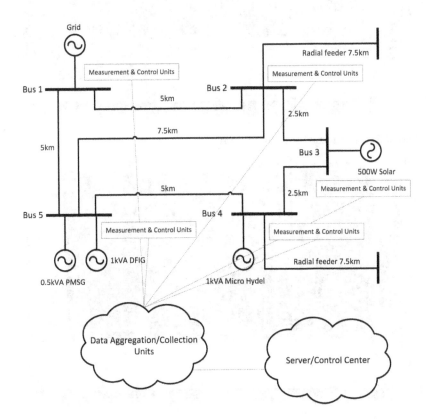

FIGURE 3.56   Communication enabled 5 bus microgrid laboratory model.

The microgrid developed in this emulator has the ability for grid-tied operation by connecting to the main power grid and enabling export or import of electricity based on local generation and demand in the microgrid. The different distributed generation schemes in this microgrid simulator include a synchronous generator acting as micro hydroelectric plant, SCIG and PMG (each coupled with a DC motor emulated as a wind turbine) acting as wind generator, and SPV panel. The storage schemes in the microgrid simulator include battery with the SPV panel and pumped storage with the micro hydro-electric plant. There are two radial feeders where consumer loads can be connected to consume the energy (Figures 3.55 and 3.56). Figures 3.57 and 3.58 show the front panel and back panel view of the laboratory scale microgrid simulator, respectively.

FIGURE 3.57   Front panel of 5 bus microgrid.

FIGURE 3.58   Back panel of 5 bus microgrid.

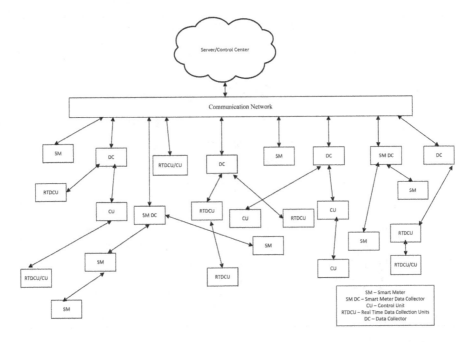

FIGURE 3.59 Communication possibilities in 5 bus microgrid.

The next set of steps explains the methodology for converting the developed hardware of microgrid simulator into a smart microgrid simulator (SMGS) (Figures 3.59 and 3.60).

- The electrical parameters to be monitored are identified and the points of measurements are decided. These measurements are to be carried out in a distributed and synchronized manner throughout the network so as to mimic a WAMS behavior in an actual power grid.
- The computation in real time to be performed using the measured data and the analysis to be followed are then decided. This helps in finalizing the computing requirements of the smart modules in microgrid.
- The data representation standards and the rates of data collection, processing and communication are finalized.
- The communication architecture, network topology and technologies that must be used for the data transfer from point of measurement to the central control center meeting the operational requirements are decided.

Real-time data collection units (RTDCU) capable of measuring real-time voltage and current, phasor computation and communication are developed and integrated on all the buses of the SMGS for aiding in automation with sensing and control actions. Eight RTDCU modules are distributed on five buses of the SMGS for real-time data collection. Design and fabrication details of RTDCU are provided in Section 4.4.3. A standalone application is used at the server side to receive, process and store

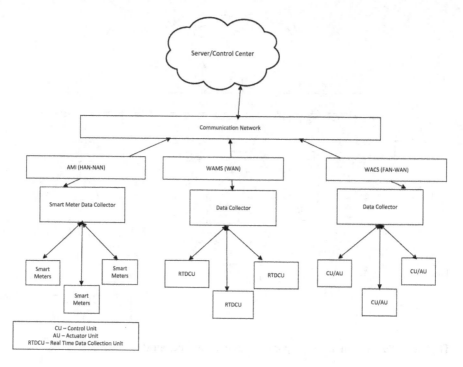

FIGURE 3.60 Communication network for 5 bus microgrid.

the data and generate any required control signal. The applications intended to be demonstrated are real-time monitoring, control and metering. Data communication for monitoring application is implemented following the IEEE C37.118.2, control using the IEEE 1815-2012 (DNP3) and metering following the IEC 62056 (DLMS/COSEM).

Communication network design can start with fixing the network topology required in terms of position of member nodes that constitute the network. The member nodes of the communication network under discussion are RTDCUs and control nodes in the microgrid as well as smart meters at the consumer premises. The decision on placement of these nodes is done based on the operational needs of the microgrid and the associated applications. The control units (CU)/RTDCU are used for real time, distributed and synchronized data collection and actuation or control action in the microgrid. The data concentrators (DC) collect the data from these nodes and pass it on to the control center. Similarly, the topology of the smart meters at consumer premises is developed considering the distributed nature of consumer premises Figures 3.61 and 3.62 shows the monitoring and control, and metering communication network on the SMGS line diagram, respectively.

The monitoring message path is RTDCU-DC-Server using the WAN to achieve WAMS. The control messages use the message path Server-DC-CU using WAN and realize WACS. The metering messages use NAN and WAN to fulfill the path SM-DC-Server. The communication system requirements and formatting for the monitoring,

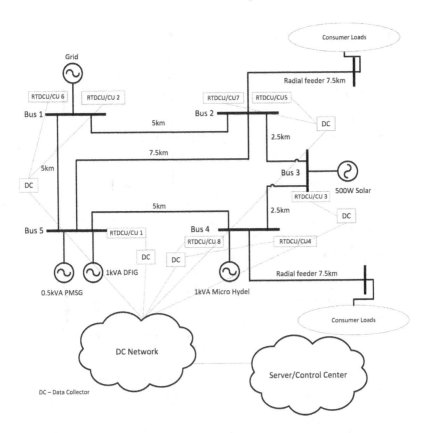

FIGURE 3.61 WAMS and WACS in 5 bus microgrid.

control and metering applications are defined in the respective standards. The constraints imposed by the statutory regulations for communication requirements also must be taken into account. These requirements can be mapped against the features and functionalities offered by various communication technologies specified in the communication technology standards. There are many eligible candidates of communication technology that match the requiremjents and hence the selection of the right one is quite tricky. This selection depends on a lot of metrics like (i) the performance capabilities of the technologies (throughput, delay, bandwidth, range, packet delivery ratio, etc.), (ii) feasibility of operation based on specified message sequence and data formatting as specified in standards, and (iii) operation and performance demand based on power system constraints and limits.

This hierarchical architecture following the required topology can be implemented in communication network simulation software packages like ns2, ns3, OMNET++, OPNET, NetSim, QualNet, etc., to evaluate the performance of the communication network with the help of performance metric parameters. The data transfer in proper standard executed in a simulation software helps to compare the performance of different communication technologies and this helps in the selection of correct technology for each link of the microgrid communication network. Here, the network for

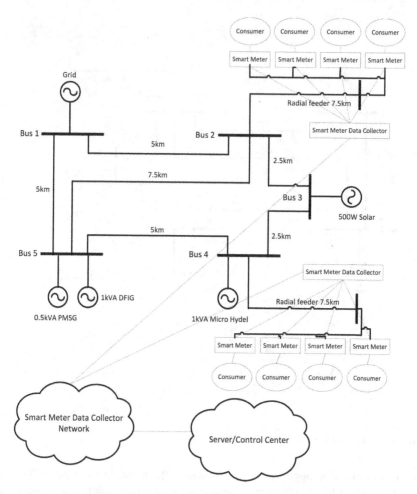

FIGURE 3.62　AMI in 5 bus microgrid.

monitoring application implements *command*, *configuration*, *data* and *header* messages as specified in the IEEE C37.118.2 standard. Similarly, the network for control application implements request and response for *select, operate, read* and *conform* and *spontaneous event* notifications as specified in IEEE 1815-2012. The metering network implements request and response for *get, set, action* and *access* messages specified in IEC 62056. The specific details of implementation including the results and its analysis are not presented here as it is not within the scope of this book. The extensive simulation with IEEE 802.3, IEEE 802.11 and IEEE 802.15.4 and link-wise performance analysis suggested that the best network for microgrid, to meet the requirements of the selected applications, would be a heterogeneous one with the backbone facilitating WAN created using IEEE 802.3 and the last miles implemented using IEEE 802.11 for monitoring and control applications and IEEE 802.15.4 for metering applications.

3.7   CHALLENGES IN COMMUNICATION NETWORK FOR MICROGRIDS

The smart microgrids will have large number of communication nodes interconnected to form networks for data exchange to realize many applications. The communication systems should be efficient, secure and available for effective data transfer between nodes. Additionally, there should be optimal utilization of available communication resources by employing methodologies to create a reliable, scalable and interoperable end-to-end communication network. With new applications introduced to improve functionalities, the number of devices increases and the corresponding smart operations adopt a decentralized approach, further increasing complexity and vulnerability of the communication network. The following are some of the main challenges of a smart grid communication system.

1. *Security*: Security is a major challenge for any smart grid communication network. To achieve high levels of security and privacy, technologies must have the capability to detect and mitigate attacks and malicious behavior. Each communication technology has its own flavor of security mechanism and in a heterogeneous system like smart grid communication network the security of the system can only be as strong as the strength of security of the weakest link. Adding heavy access control and encryption techniques are challenging as this would increase the computational complexity and would adversely affect the latency. Also, different classes of messages need different levels of security.

2. *Compression*: The amount of data generated and transmitted for various smart grid applications is huge and could easily exhaust the available resources of the existing communication infrastructure. This would force to use compression techniques like coding schemes. The challenge in this is the computational load that would be added to the node and also the latency that would be introduced.

3. *Unavailability of spectrum*: A large number of communication nodes also implies, particularly in wireless communications, that the allocated spectrum for communication may get exhausted. Using unlicensed spectrum has the challenge associated with its access and communication blockage due to interference, both negatively affecting the quality of service. Use of a cognitive system is a solution to this problem as it could increase the spectrum efficiency, as spectrum sensing in a wide frequency range is possible in cognitive networks. The computational overhead, latency and dynamic network configuration needs are the associated challenges.

4. *Redundancy*: As smart grid is a critical infrastructure, poor performance or failure of tightly integrated communication network can collapse the grid, redundancy in terms of nodes and communication links must be there for the network path to deal with critical data. The cost involved and the synchronization to dynamically switch to active mode from hot standby are the challenges involved here.

5. *Interference management schemes*: Interference management schemes are to manage the harsh environment or bad links. The main sources of interference include electricity flows, use of multiple communication technologies and the density of communicating devices. The solution is to select appropriate interference-free channel resources using techniques like relaying and cooperative beamforming. Again challenges in the form of computational complexity and latency are the issues here.

6. *Resource constraint nodes*: Smart grid is a large and complex system interconnecting enormous number of devices with significantly diverse computation and communication capabilities. Addition of new algorithms and techniques for performance improvement is always a challenge as it would mean more computation power at disposal, more energy, redesign of software stack to ensure deterministic operation, etc., to be accomplished at low or no additional cost. The problem multiplies many times over when this is to be incorporated in many nodes of the network as a smart grid communication network typically is a collection of heterogeneous nodes and technologies.

7. *Interoperability*: All the applications of smart grid, as well as technologies that implement these applications are having their own standards. Having a synchronized operation of all these devices and technologies, under this scenario, is the biggest challenge. Interoperability requirements exist at various levels in smart grid communication system including protocols, standards, nodes, network operations, and applications.

BIBLIOGRAPHY

1. Saleh, M., Esa, Y., Hariri, M.E. and Mohamed, A., Impact of information and communication technology limitations on microgrid operation. *Energies*, vol. 12, no. 15, pp. 1–24, 2019.
2. Marzal, S., Salas, R., González-Medina, R., Garcerá, G. and Figueres, E., Current challenges and future trends in the field of communication architectures for microgrids. *Renewable and Sustainable Energy Reviews*, vol. 82, Part 3, pp. 3610–3622, 2018.
3. Dib, L.D.M.B.A., Fernandes, V., de L. Filomeno, M. and Ribeiro, M.V., Hybrid PLC/wireless communication for smart grids and internet of things applications. *IEEE Internet of Things Journal*, vol. 5, no. 2, pp. 655–667, 2018.
4. Khan, F., Rehman, A.U., Arif, M., Aftab, M. and Jadoon, B.K., A survey of communication technologies for smart grid connectivity. *International Conference on Computing, Electronic and Electrical Engineering (ICE Cube)*, Quetta, 2016, pp. 256–261.
5. Mulla, A., Baviskar, J., Khare, S. and Kazi, F., The wireless technologies for smart grid communication: A review. *Fifth International Conference on Communication Systems and Network Technologies*, Gwalior, 2015, pp. 442–447.
6. Kuzlu, M., Pipattanasomporn, M. and Rahman, S., Communication network requirements for major smart grid applications in HAN, NAN and WAN. *Computer Networks*, vol. 67, pp. 74–88, 2014.
7. Fan, Z., Kulkarni, P., Gormus, S., Efthymiou, C., Kalogridis, G., Sooriyabandara, M., Zhu, Z., Lambotharan, S. and Chin, W.H., Smart grid communications: Overview of research challenges, solutions, and standardization activities. *Communications Surveys & Tutorials, IEEE*, vol. 15, no. 1, pp. 21–38, 2013.

8. Güngör, V.C., Sahin, D., Kocak, T., Ergut, S., Buccella, C., Cecati, C. and Hancke, G.P., Smart grid technologies: Communication technologies and standards. *IEEE Transactions on Industrial Informatics*, vol. 7, no. 4, pp. 529–539, 2011.

9. Tan, S.K., Sooriyabandara, M. and Fan, Z., M2M communications in the smart grid: Applications, standards, enabling technologies, and research challenges. *International Journal of Digital Multimedia Broadcasting*, vol. 2011, pp. 1–8, 2011.

10. Department of Energy, Communications Requirements of Smart Grid Technologies, Washigton, DC, October 5, 2010.

11. Communication networks and systems for power utility automation-All parts, IEC Standard 61850, 2019.

12. IEEE Standard for Ethernet. In *IEEE Std 802.3-2018 (Revision of IEEE Std 802.3-2015)*, pp. 1–5600, August 31, 2018.

13. IEEE Standard for Information technology—Telecommunications and information exchange between systems Local and metropolitan area networks—Specific requirements – Part 11: Wireless LAN Medium Access Control (MAC) and Physical Layer (PHY) Specifications. In *IEEE Std 802.11-2016 (Revision of IEEE Std 802.11-2012)*, pp. 1–3534, December 14, 2016.

14. IEEE Standard for Low-Rate Wireless Networks. In *IEEE Std 802.15.4-2015 (Revision of IEEE Std 802.15.4-2011)*, pp. 1–709, April 22, 2016.

15. IEEE Guide for Phasor Data Concentrator Requirements for Power System Protection, Control, and Monitoring. In *IEEE Std C37.244-2013*, pp. 1–65, May 10, 2013.

16. IEEE Standard for Electric Power Systems Communications-Distributed Network Protocol (DNP3). In *IEEE Std 1815-2012 (Revision of IEEE Std 1815-2010)*, pp. 1–821, October 10, 2012.

17. IEEE Standard for Synchrophasor Data Transfer for Power Systems. In *IEEE Std C37.118.2-2011 (Revision of IEEE Std C37.118-2005)*, pp. 1–53, December 28, 2011.

18. *IEEE Guide for Smart Grid Interoperability of Energy Technology and Information Technology Operation with the Electric Power System (EPS), End-Use Applications, and Loads*, IEEE Std 2030-2011, September 5, 2011.

19. *Electricity Metering – Data Exchange for Meter Reading, Tariff and Load Control – Part 21: Direct Local Data Exchange*, IEC Standard 62056-21, 2002.

20. Bian, D., Kuzlu, M., Pipattanasomporn, M., Rahman, S. and Shi, D., Performance evaluation of communication technologies and network structure for smart grid applications. *IET Communications*, vol. 13, no. 8, pp. 1025–1033, 2019.

21. Pathirikkat, G., Mallikajuna, B. and Bharata Reddy, M.J., Recent Trends on Performance Analysis of Latency on Wide Area Technologies in Smart Grids. *20th National Power Systems Conference (NPSC)*, Tiruchirappalli, India, 2018, pp. 1–6.

22. Hiew, Y., Aripin, N.M. and Din, N.M., Performance of cognitive smart grid communication in home area network. *2014 IEEE 2nd International Symposium on Telecommunication Technologies (ISTT)*, Langkawi, 2014, pp. 417–422.

23. Kuzlu, M. and Pipattanasomporn, M., Assessment of communication technologies and network requirements for different smart grid applications. *Innovative Smart Grid Technologies, IEEE PES*, pp. 1–6, February 24–27, 2013.

24. Dong, Y., Cai, Z., Yu, M. and Sturer, M., Modeling and simulation of the communication networks in smart grid. *Systems, Man, and Cybernetics, IEEE International Conference on*, pp. 2658–2663, October 9–12, 2011.

25. Sood, V.K., Fischer, D., Eklund, J.M. and Brown, T., Developing a communication infrastructure for the smart grid. *Electrical Power & Energy Conference, IEEE*, pp. 1–7, October 22–23, 2009.

26. Emmanuel, M., Seah, W.K.G. and Rayudu, R., Communication architecture for smart grid applications. *2018 IEEE Symposium on Computers and Communications (ISCC)*, Natal, 2018, pp. 00746–00751.

27. Kabalci, E., Kabalci, Y., *Smart Grids and Their Communication Systems*, 1st edition, Singapore, Springer, 2019.
28. Leccese, F., An overview on IEEE Std 2030. *2012 11th International Conference on Environment and Electrical Engineering*, Venice, pp. 340–345, 2012.
29. Ali, I., Aftab, M.A., Hussain, S.M.S., Performance comparison of IEC 61850-90-5 and IEEE C37.118.2 based wide area PMU communication networks. Journal of Modern Power Systems and Clean Energy, vol. 4, no. 3, pp. 487–495, 2016

4 ICT Application of DSM

S. Nithin

CONTENTS

4.1 INTRODUCTION

Maximizing RE utilization is the ultimate goal for utilities in realizing the green energy concept. However, the uncertainties in RE generation force utilities to employ RE curtailment which weakens green energy initiatives. The scope of demand side management (DSM) strategies to handle the stochastic nature of demand and generation has attracted smart energy research efforts. The convergence of energy systems and ICT to enable smart operations has led to the development of smart versions of DSM like demand response (DR) and demand dispatch (DD). The ICT applications like WAMPAC and AMI have been discussed in the previous chapter. The DSM is yet another ICT application for smart grids.

This chapter introduces the DSM schemes of DR and DD and then reports in detail the development of a DD framework. Testing of the DD algorithm on the smart microgrid emulator as well as on a DC microgrid is also presented.

4.2 DEMAND RESPONSE

The demand for electric power is ever increasing and utilities are finding it difficult to manage the fluctuations in demand. The concept of distributed generation promotes interconnection of more RE sources, however the intermittency and variability in RE aggravates the existing demand-supply mismatch. The DSM strategies address the issues of demand-supply mismatch. DSM gained popularity owing to its

efficacy in shaping the demand curve. Demand response is one such load management technique put in practice by the grid operator in order to regulate the demand
as and when required. DR programs were originally proposed to curtail some of
the bulk loads during peak times, via telephone calls. DR is usually associated with
only a few consumers and is used infrequently. In a DR contract with the utility the
consumer entrusts the utility to disconnect the contracted load when the latter feels
the need. DR is an effective tool; however, its success depends on the identification
of controllable loads. Utilities assign this task to *demand side response aggregators* who collect and control DR loads upon instruction from the utilities. The DR
switches, which receive control signals from utilities, are deployed on the consumer
side and to control the loads. The DR programs could be primarily divided into two
categories: (a) Dispatchable and (b) non-dispatchable/time-based. Detailed classification is given in Figure 4.1.

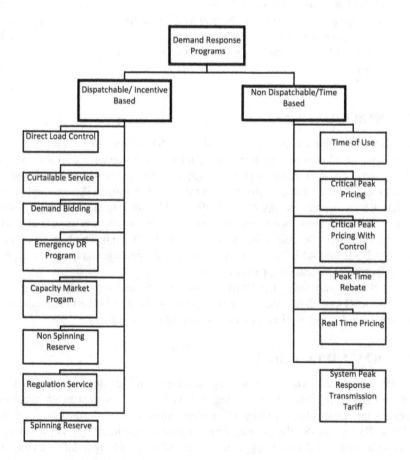

FIGURE 4.1   DR classification.

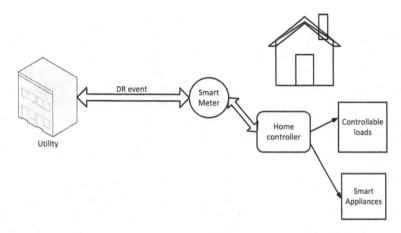

FIGURE 4.2   DSM control.

4.2.1   Communication and Control

The communication infrastructure of AMI aids the utility in deployment of DR operations. The introduction of smart appliances under IoT paradigm and smart home facilitates DSM by utilities. Typical DR communication is depicted in Figure 4.2.

The DSM strategy of DR was introduced by FERC in 2006 to handle demand during peak hours of consumption. The success of the early versions of DR motivated the development of more sophisticated versions, through which behind-the-meter resources – such as loads, RE generation, diesel generators, etc., are controlled to maintain supply-demand synergy.

4.3   DEMAND DISPATCH

The recently evolving DSM strategies can handle the stochastic nature of both demand and generation. The convergence of power systems and ICT to enable smarter operations lead to the development of the advanced version of DR, called demand dispatch (DD). Advanced communication technologies helped to extend DR schemes as DD, where the utility can aggregate consumer loads and dispatch them in tandem with the generation, thus forcing demand to follow generation. DD uses the concept of dispatchable loads (DLs). The DLs are loads that can deviate from their normal consumption pattern without affecting their operational constraints. For the success of DD, the utility needs to identify, aggregate and precisely control DLs. Traditional power system operation (i.e., where generation follows demand) is no longer effective in the context of variable RE. The availability and controllability of DLs are of prime concern for DD. The contract between the utility and the consumers ensures seamless control of DLs whenever needed by the utility.

Multiple demand side aggregators are needed for DD operations in a large utility grid. Since DD is a new management strategy, the operational framework and the various entities involved in operations are yet to be identified and standardized.

The DD requires real-time communication between appliances/loads and utility control centers. Latency of communication within the grid is of great importance as some loads should be dispatched first in cases of power system urgency. Internet is a promising medium through which DD could be achieved, as the infrastructure already exists. The utility can implement a dynamic tariff plan in conjunction with DD so that the change in price for electric power generation is reflected to the consumers as well. Dynamic tariffs will promote voluntary involvement of consumers in DD schemes. Moreover, the consumer can select the operation pattern of loads in such a way as to minimize his total consumption.

The major challenge to be addressed in DD is this customer participation aspect. The motivating factors for participating in a DD or DR program are the associated incentives. For DD/DR schemes to succeed, the consumer should be aware of dynamic tariff, time ratings and their own consumption patterns. To tackle these issues, the DD/DR schemes should be automated to incorporate consumer preferences in load connection/disconnection decisions while maintaining synergy with grid operations. The utility has to aggregate and precisely control dispatchable loads to nullify the fluctuations induced by variable RE, like wind farms. The optimum selection of dispatchable loads based on spatial constraints is crucial as the changes in demand would affect voltage stability and power flow in the grid. The utility has to select appropriate communication technology and security measures conforming to various standards like ISO 27002, NIST SP800-53, NERC CI, ISA.

4.3.1 ROLE OF AGGREGATORS

The utility operators can employ DD through direct load control; however, it is desirable for huge power systems to use aggregator services. The architecture of DD is evolving; therefore, DD architecture, which is similar to that used for DR, is discussed here. The architecture comprises of three layers: The first layer involves the utility, demand dispatch provider (DDP) and independent power producers (IPP); these entities interact in the electricity market. The IPPs forecast their net generation on a day-ahead basis and tender their price accordingly. Utilities in the first layer bid for IPP forecasted generation and request DDP service in anticipation of forecasting errors. The DDP relies on several demand dispatch aggregators (DDA) for collecting information on DL from consumers, controlling DL and providing incentives to consumers for their participation in DD. The DDAs can be broadly classified into: (1) Commercial DDA, (2) industrial DDA, and (3) residential DDA. The role played by each DDA is summarized in Table 4.1.

The second layer of the architecture consists of the numerous DDAs performing DD operation based on the requirements of DDP. Dispatchable loads under the control of commercial, industrial, and residential DDAs form the third layer of the architecture.

The DDP requests DDA to collect DL information and announces the dispatch capacity required for each time slot of operation. DL owners belonging to each sector enroll their DLs on a day-ahead basis against the incentives offered by DDAs. Finally, DDAs

TABLE 4.1

Types of DD Aggregators

Type of DDA	Associated Dispatchable loads
Commercial DDA	Shopping malls, movie theatres, etc., with backup power supply, EV charging stations, diesel generators, etc.
Industrial DDA	Water pumping stations, flexible demand in processing plants, etc.
Residential DDA	Water heaters, pool pumps, irrigation system, residential EV charging, battery packs, small scale generation system, etc.

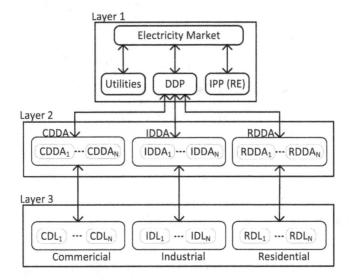

FIGURE 4.3 DD architecture CDDA: Commercial DDA; IDDA: Industrial DDA; RDDA: Residential DDA; CDL: Commercial DL; IDL: Industrial; RDL: Residential DL.

submit bids to DDP for service cost against dispatch capacity. The DD architecture is depicted in Figure 4.3 and information flow in the architecture is depicted in Figure 4.4.

Each aggregator announces the extent of demand they can change (increase/decrease) for each operational time slot on a day-ahead basis. Aggregators do this on the basis of DLs enrolled by the consumers and priorities assigned by the consumers for each DL. Aggregators collect such information through an online portal, where consumers can enroll their DLs, set minimum operational time of each DL in a day and assign a priority for each DL. It is required that aggregators are equipped with such data, have assigned a priority rating for each DL considering consumer preferences and have announced the service charge for each time slot.

The incentives offered to consumers vary across the day and hence the service costs of aggregators vary with time. The service cost per kW for each operational

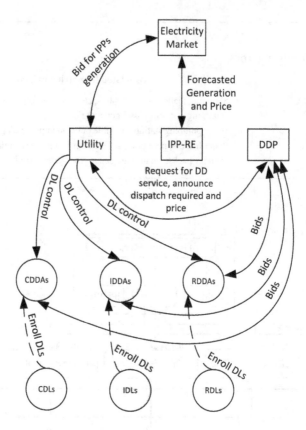

FIGURE 4.4   Information flow in DD architecture.

time slot is also announced by the aggregators on a day-ahead basis. This variable incentive scheme attracts more consumers to participate in DD operations. The aggregators benefit as they are paid by the utility for the DD service. The utility has to compensate for the services offered by the aggregator and this further increases the cost of utility operation. At high levels of RE penetration, the goal of the utility is to dispatch more RE and maximize net revenue. The utility deploys DD to absorb the variable RE power injected into its system.

The role of aggregators is to facilitate DD and thereby maximize their profit, without sacrificing consumer priorities. A single aggregator might not be sufficient to facilitate DD in a large grid; when multiple aggregators are present, the utility needs to precisely estimate the volume of demand each aggregator has to dispatch as depicted in Figure 4.5. Among the numerous demand-side aggregators existing in the grid, the utility needs to judiciously assign a dispatch share, so that each aggregator can adjust DLs under its control, thus leading to optimum DD. This process is termed aggregator dispatch share allocation (ADSA). The ADSA should consider grid constraints with respect to voltage stability and feeder overloading.

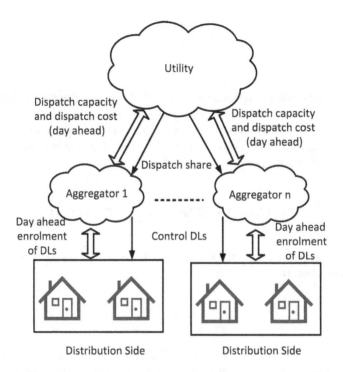

FIGURE 4.5   Schematic of ADSA.

4.3.2   ADSA FORMULATION

The potential of DD to reshape the demand curve in real time is utilized here to adjust to changes in RE generation. A utility can employ DD to absorb RE generation directly or as a tool to account for unscheduled RE generation. Though the operational aspects are similar in both scenarios, the key factor is the availability of DLs in large numbers. It is assumed here that DD is assigned with the former task while DLs are available in surplus numbers from various DDAs.

Consider a utility grid with n buses where n_{pq} buses feed loads. Let P_R be the instantaneous RE penetration at the time instant of t and ΔP_R be the change in RE injection with respect to the previous instant. This is expressed as,

$$\Delta P_R(t) = P_R(t) - P_R(t-1) \tag{4.1}$$

where $P_R(t)$ is the RE injection at the instant of t and $P_R(t-1)$ is the injection at the previous instant. It is assumed that the aggregator has the net DL capacity at each of the n_{pq} buses represented by $P_{\text{dispatch}}(i) \epsilon \{1,2\ldots n_{pq}\}$, and let P_{DD} be the aggregate dispatch needed on each of the n_{pq} buses at that instant. P_{DD} of all the buses, when aggregated, equal ΔP_R; that means,

$$\Delta P_R = \sum_{i=1}^{n_{pq}} P_{DD}(i) \tag{4.2}$$

where $0 \leq P_{DD}(i) \leq P_{dispatch}(i)$.

Let the utility procure RE at a rate of Rs. C_R/kWh, and pay Rs. C_{DD}/kWh as dispatch service cost to m number of aggregators. Let the utility sell energy to consumers at a rate of Rs. C_L/kWh, where $C_R < C_L$. The incentive given to the consumers will be Rs. C_I/kWh. These power and cash flows are depicted in Figure 4.6. The net income (*NI*) of the utility can be expressed as,

$$NI = \left(P_{DD}C_L\right) - \left\{\left(\Delta P_R C_R\right) + \sum_{i=1}^{n_{pq}} P_{DD}(i)\, C_{DD}(i)\right\} \tag{4.3}$$

The utility has to maximize the net revenue through DD; the corresponding maximization function is,

$$\max_{P_{DD}} NI \tag{4.4}$$

subject to constraints, $C = V_{\min}(i) \leq V(i) \leq V_{\max}(i)$, $\forall\ i \in \{1, 2, \dots, n\}$ where $V_{\min}$ and $V_{\max}$ are the lower and upper limits respectively of the bus voltages in per unit (p.u.).

Addition of extra demand on the grid has adverse effects on bus voltages and hence this is modeled as a penalty factor P, which is a vector of size n and can be expressed as,

$$P(i) = \begin{cases} \text{if } V(i) > V_{\max}(i),\ P(i) \text{ is } + \text{ve} \\ \text{else if } V(i) < V_{\min}(i),\ P(i) \text{ is } - \text{ve}, \forall\ i \in \{1, 2, \dots, n\} \\ \qquad \text{else, } P(i) = 0 \end{cases} \tag{4.5}$$

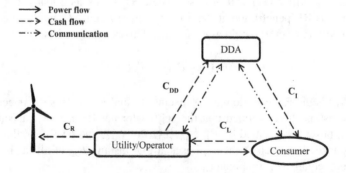

$\longrightarrow$ Power flow
$\dashrightarrow$ Cash flow
$-\cdots\rightarrow$ Communication

FIGURE 4.6 Power flow, cash flow, and communication/control flow of DD operation.

The addition/removal of demand to maintain voltage within these limits is a constraint and hence P is used as a factor in calculating the cost. Penalty values are assigned as a function of ΔP_R, so that voltages at all buses are maintained within the limits by controlling the dispatched load. This is modeled as a cost, by multiplying the quantum of demand to be modified on the respective bus by a voltage penalty of C_{VV}. Here C_V represents the aggregate cost of voltage deviation on all the buses as,

$$C_V = \sum_{i=1}^{n} P(i) C_{VV} \qquad (4.6)$$

The line flow limits are determined through Newton-Raphson (NR) iterative load flow analysis. The non-convergence of NR load flow is also modeled as a higher cost denoted as C_{LF},

$$C_{LF} = \left\{ \begin{array}{c} 0, \text{ if load flow converges} \\ 10{,}000, \text{ if load flow does not converge} \end{array} \right\} \qquad (4.7)$$

From (3.3), (3.5), (3.6) and (3.7) the fitness F for each particle can be evaluated as,

$$F = P_{DD} C_L - \left[\Delta P_R C_R + \sum_{i=1}^{n_{pq}} P_{DD}(i) C_{DD}(i) + C_v + C_{LF} \right] \qquad (4.8)$$

The objective function is given in (4.4), fitness function is in (4.8) and constraints are given in (4.5) and (4.7). The particle swarm optimization (PSO) method is adopted here. Particles are generated in increments of P_{DD} at each bus. Their fitness is evaluated using (4.8) through NR load flow. Global best and personal best fitness values of particles are computed and particles are updated based on velocity. The PSO retire particles from the iteration once tolerance has been met; final particles indicate the demand to be dispatched on each bus and so the aggregators in charge of buses can control DLs, respectively.

4.3.3 SELECTION OF LOADS

The DD framework discussed here relies on DDAs for identification and aggregation of DLs from among commercial, industrial and residential consumers. This is a cumbersome task as such participants will be copious. An efficient way to address this is to rely on energy management systems (EMS) to be used in all the sectors. Home-EMS, building area network under smart grid paradigm and industrial IoT (IIoT) concepts could be utilized for the development of such EMS. Such an attempt is reported here to develop an EMS architecture irrespective of the operational sector. DLs can be smart appliances with built-in intelligence and communication. However, a low-cost solution can be adopted to transform existing appliances to be compatible with smart operations. Such solution may include a load control unit (LCU), a central EMS unit and necessary communication links. The schematic of EMS is shown in Figure 4.7.

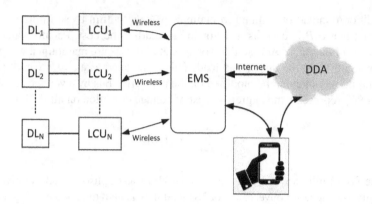

FIGURE 4.7   Schematic of EMS.

The EMS depicted in Figure 4.7 ensures the enrollment of DLs for DD via mobile phone application. The mobile phone application establishes a bidirectional medium through which consumers and DDA can interact. Such interactions result in announcement of incentives and enrollment of DLs in a day-ahead basis. DDAs make use of these data and formulate their bids. The EMS could be equipped with smart operations devoid of DD, too. The suggested system will also enable remote control of non-DL, as well. The LCU modules could be equipped with multiple communication capabilities depending on the deployment requirements. LCU modules are capable of power measurement and control of DLs. Once the user registers a DL, its power consumption is fed to DDA by the LCU. LCUs, depicted in Figure 4.8, keep track of the power consumption of DLs and update the state of operation in the DDA server/cloud. DDAs can collect DL information and prioritize these based on consumer inputs such as minimum operational hours, DL schedule, etc.

Several criteria indices of DL prioritization, suggested for DR in the literature, can be used by DDAs in the case of DD as well. These include appliance priority index, appliance flexibility index, appliance satisfaction index, power similarity index, high power consumption index, etc. Furthermore, DDAs can inform each EMS unit of the amount of DLs to be connected/disconnected during the DD operation. EMS will

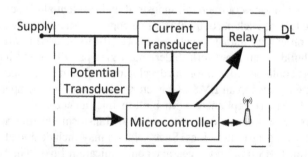

FIGURE 4.8   Schematic of LCU.

follow the instructions from DDA, and if the dispatch share is less than the available DL capacity, then EMS can prioritize DL operation considering consumer preferences as well.

The load selection process by the aggregator is a combinatorial optimization problem, where the aggregator should find DL matching the ADS. In fact, the DL selection problem exhibits similarities to 0–1 Knapsack problem, where a knapsack/backpack is to be filled with items to maximize profit without exceeding the knapsack's weight carrying capacity. Each item available for selection has a respective weight and value; the algorithm should select the best items so that their combined values are the highest and yet are well within the weight capacity of knapsack. Here, DLs available for DD are considered to be *items*, the rated power of each DL is taken as its *weight* and the priority assigned to each DL by the aggregator is taken as *item value*. The objective is to select high priority (value) DLs where the aggregate of selected DLs (weights) equals the required ADS. Selection of DLs modeled as a 0–1 knapsack problem, is given in equations 4.9 and 4.10, and solved through dynamic programming. Assuming *nl* number of DLs, ADS is taken as the knapsack's capacity, DL ratings are taken as *item weights* and priority assigned to each DL is taken as *item value*.

$$\text{Maximize} \sum_{i=1}^{nl} \text{DL}_{\text{Priority}_i} S_i \tag{4.9}$$

subject to

$$\sum_{i=1}^{nl} (\text{DL}_{\text{Rating}_i} S_i) \leq \text{ADS and } S_i \in \{0,1\} \tag{4.10}$$

where $\text{DL}_{\text{Priority}_i}$ is the priority assigned to each DL, $\text{DL}_{\text{rating}_i}$ is the power rating of each DL and S_i denotes if a DL is selected or not. Dynamic programming divides a problem into sub problems, solves each of these and combines the solutions. The recursion formula for the knapsack is,

$$f(k,g) = \begin{cases} f(k-1,g) \text{ if } W_k > g \\ \max\{V_k + f(k-1,g-W_k), f(k-1,g) \text{ if } W_k < g \wedge k > 0 \end{cases} \tag{4.11}$$

where $k \in \{1, 2, \ldots, nl\}$, $g = \text{ADS}$, $V_k = $ net value considering kth DL, $W_k = $ power rating of kth DL.

Equation (4.9) represents the objective function, where DLs with high priorities are selected. The selection of high priority DLs should be performed with a constraint that the aggregate of selected DLs' power consumption should not be higher than the required power to be dispatched; hence it is subjected to the constraint given in (4.10). The recursive operation in (4.11) checks each DL's priority and rated power based on dynamic programming, then decides to select or reject it based on (4.9) and (4.10).

4.4 DD IMPLEMENTATION ON MICROGRIDS

The co-simulation of both power system and communication network is a necessity for research on smart grids. Few software packages possess co-simulation capabilities; so it still remains a cumbersome task to manage the time constraints. Hardware in loop (HIL) systems and real-time digital simulators (RTDS) offer better controllability, but their use is restricted due to the high cost involved. Most of the test beds are designed for specific power system research and hence operated at high voltage and power. These in turn motivated development of the low-cost generic hardware simulator suitable for studies on microgrids – the smart microgrid simulator (SMGS) which is discussed in Section 3.6 and shown in Figures 3.59 and 3.60.

4.4.1 SMART MICROGRID SIMULATOR & WT EMULATOR

In this section, the hardware simulator is used for the validation of DD along with a wind turbine (WT) emulator. The WT emulator scheme is shown in Figure 4.9; the DC motor in series with the resistor emulates the wind turbine. The characteristics of the WT emulator are shown in Figure 4.10, where C_p is the power coefficient, which is the ratio of its mechanical power output to the wind power input, and λ is its tip speed ratio, equal to the ratio of the speed at the tip of the turbine blade relative to the wind speed (as defined by equation 2.97 in Chapter 2).

To employ smart operations in the microgrid, real-time data collection is necessary, hence Schneider EM 6436 meters are connected in the laboratory scale microgrid simulator to acquire real-time values of bus voltages, frequency and power transfers. The meters provide data every second via Modbus RTU protocol. Raspberry Pi 3 (Rpi) model is configured to act as the Modbus master to collect data from the meters through a USB to RS-485 convertor. The python library (pymod) is used in Rpi for Modbus implementation. The number of demand side response aggregators in the microgrid is two, as the system involves only two radial feeders for distribution. Each aggregator has the data and control of the respective DL in his command.

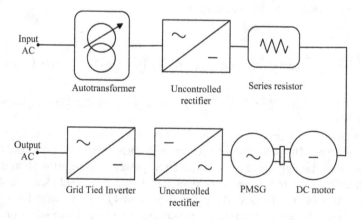

FIGURE 4.9   WT emulator.

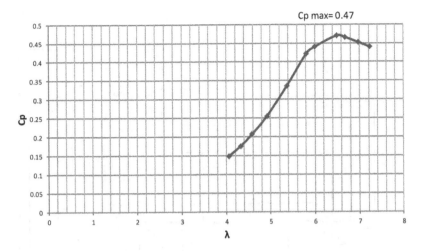

FIGURE 4.10 $C_p - \lambda$ plot of the WT emulator developed in the laboratory.

A few DL are located such that these can be connected to (and disconnected from) the radial feeders by use of LCU as and when requested by the DD application in the cloud. The DD application monitors the injected wind power and the system frequency and provides control signals to LCU to connect/disconnect loads. LCU are developed based on NodeMCU and these communicate over Wi-Fi. The DD application also keeps track of the amount of load added or removed. Meter data are updated to the Ubidots cloud via Internet of Things (IoT) protocols, whereas LCU also use the same cloud infrastructure. The experimental set up is shown in the schematic diagram of Figure 4.11.

The laboratory scale implementation of DD used the SMGS discussed earlier. The microgrid is autonomous and it has a WTG connected to one of the buses, besides a micro hydroelectric power plant of 1 kVA as the main source. The micro hydroelectric power plant is emulated using a DC motor driven alternator of 1 kVA and WTG is emulated as discussed previously. The SMGS used for the study is shown in Figure 4.12.

4.4.2   VALIDATION OF DD ON SMGS

Given a fixed generation schedule for the alternator, with fixed loads placed on the radial feeders, the grid frequency is maintained at 50 Hz without power injection from WTG initially. The laboratory-emulated WTG is then interfaced to the grid through a grid-tied inverter following which the power input to the WT emulator is varied by varying its DC supply voltage; thus, generation at different wind speeds is emulated. The alternator and the WTG were sharing the fixed demand on the SMGS during this operation. Figure 4.13 depicts the power exports by both the alternator and the WTG during SMGS operation. While the alternator delivers a constant power to the SMGS that has no load variation, frequency excursions are observed on SMGS due to the varying WTG generation as shown in Figure 4.14.

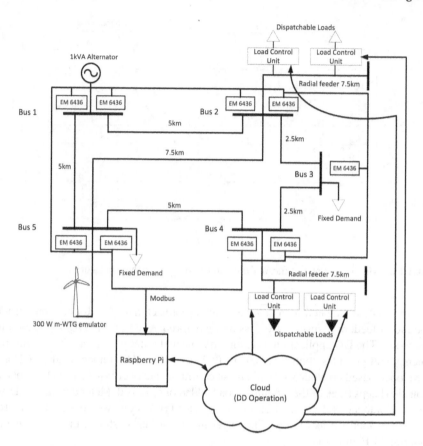

FIGURE 4.11 Experiment setup for DD implementation on SMGS.

FIGURE 4.12 SMGS used for the study.

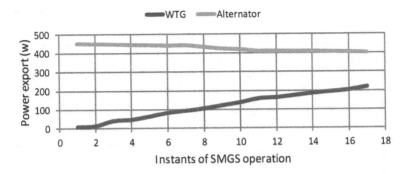

FIGURE 4.13 Power exported by WTG and alternator during SMGS operation.

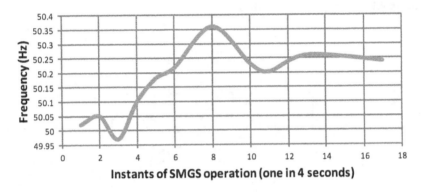

FIGURE 4.14 Frequency excursions on SMGS.

The power injected by the WTG is continuously monitored and based on that the decision is made as to the number of DL to be connected/disconnected to maintain the frequency within a narrow band around 50 Hz is made by the DD application. The command to connect or disconnect DL is sent from the cloud to LCU, through *message queuing telemetry transport* (MQTT) protocol and then the LCU interprets the data and controls DL. The proposed PSO-based ADSA algorithm is implemented in Rpi which takes the measured data from the cloud and computes the optimum share of aggregators. The algorithm performs load selection for aggregators based on dynamic programming.

The control strategy first explores the possibility of frequency correction using the injected wind power and then deploys appropriate management of DL to smooth frequency fluctuations. A set of 10 DL rated between 3 and 60 W are chosen for the task. The priority of each DL is assumed to be assigned by the aggregator prior to the operation. The test was repeated over a range of DC supply voltage, depicting corresponding range of wind speed. The DD strategy was found to be capable of dispatching DL with respect to change in WTG export. Figure 4.15 depicts the DD operation and DL dispatched by each aggregator at each of the instants at an interval of 4 s. The DD algorithm considers frequency correction first; that means, if the frequency can be improved with the help of WTG export, then DL are not dispatched.

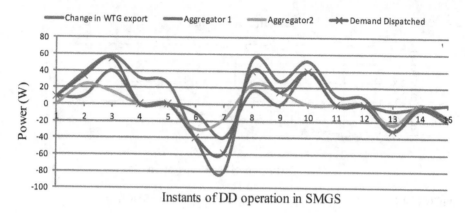

FIGURE 4.15   DD operation on SMGS.

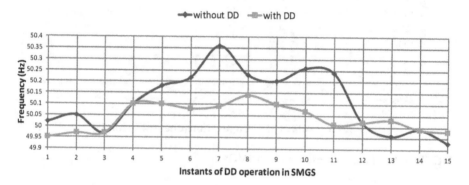

FIGURE 4.16   Frequency variations in SMGS with and without DD.

The algorithm performs DL dispatch if frequency is above the nominal value and WTG export is positive. The negative regions depicted in Figure 4.15 are the decrease in WTG export, during instances when DL are removed. The SMGS frequency with DD operation is depicted in Figure 4.16.

It can be noted from Figure 4.16 that the proposed DD strategy is capable of maintaining frequency within the permissible limits of 49.7–50.1 Hz as per the Indian grid code. However, excursions are observed during instants 3 and 4, which is due to the non-availability of DL of the required size at the respective instants. The DL available for dispatch were higher than the wind penetration. At the eighth instant, the algorithm determined the optimum amount of DL to be dispatched subject to the constraint of bus voltage violation and that led to fewer DL dispatch resulting in a frequency excursion, albeit a small one. The time consumed for the entire operation, i.e., meter data collection, PSO-based optimum dispatch selection, and knapsack-based DL selection – was approximately 32 ms, and hence suitable for real world implementation.

4.4.3 DD Implementation on DC Microgrid

The validation of DD on SMGS has been carried out as discussed in Section 4.4.2. Nevertheless, the influence of DD on a real system in the field is of more interest. A real DC microgrid, that involves a micro WTG (m-WTG), installed on the rooftop renewable energy laboratory of Amrita School of Engineering, Coimbatore, is chosen for the field testing of DD. The DD strategy is then tailored to suit the operational constraints of this DC microgrid. The m-WTG has a power rating of 100 W at a wind speed of 10 m/s. The microgrid is to operate in islanded mode, delivering power to local loads. It has battery support to serve DC loads at a nominal voltage of 12 V. The DC microgrid is shown in Figure 4.17.

The task assigned to DD here is to regulate the voltage of the DC microgrid within permissible limits of +7.5% and −5%; it means a range from 11.4 to 12.9 V. Implementation of DD requires real-time synchronized measurements on the microgrid for which the real-time data collection units (RTDCU) have been used. The role of RTDCU on SMGS has been discussed earlier in Section 3.6. One of the RTDCU units is shown in Figure 4.18. The RTDCU is capable of simultaneously sampling DC voltage and current and then computing power consumption. The RTDCU is time synchronized with the help of a common triggering pulse and two such units monitor the m-WTG generation, bus voltage, local demand, and change in demand during DD. The RTDCU is equipped with ZigBee to communicate these measurements to a centralized processing unit where decisions are made. Figure 4.19 depicts the DD strategy on DC microgrid. The regular demand on the microgrid is 20 W.

Each DL is connected to the microgrid through LCU. The LCU is equipped with a ZigBee transceiver and a microcontroller to interpret the data and control signals. The LCU is capable of power measurement as well.

The grid voltage and wind power injection are monitored in real time and DL are controlled to regulate the grid voltage. The priority of each DL is pre-assigned for the task. A total of 8 DL are considered for the implementation of DD on the DC microgrid. The power rating of DL varies from 1 to 5 W.

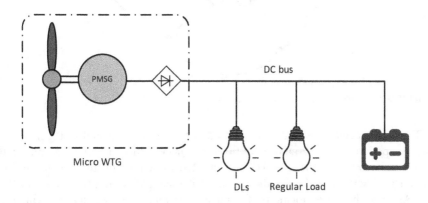

FIGURE 4.17 DC microgrid selected for the test.

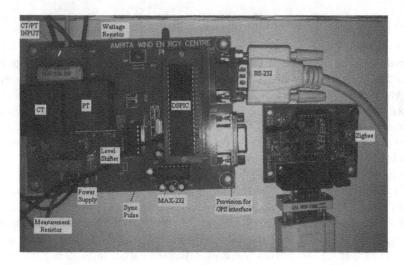

FIGURE 4.18   Real-time data collection unit.

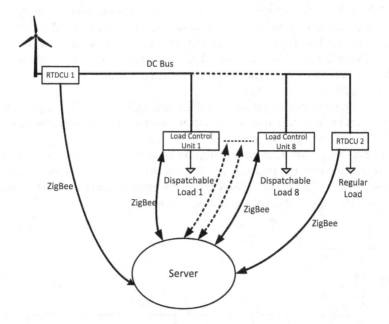

FIGURE 4.19   DD strategy on DC microgrid.

The wind speed measured and recorded at the site is presented in Figure 4.20. The test period is a little more than 90 min. The variation in wind was reflected in power injection by the m-WTG and that led to voltage fluctuation in the DC microgrid. Figure 4.21 depicts these voltage fluctuations during the period of the test. As seen, there are violations of voltage constraint at many instants during the period of observation.

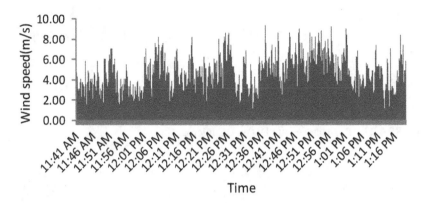

FIGURE 4.20   Wind speed recorded (every second) in the month of June 2018 at the test site.

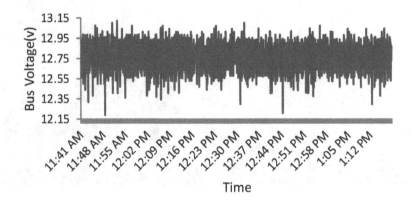

FIGURE 4.21   Voltage fluctuations during wind power injection.

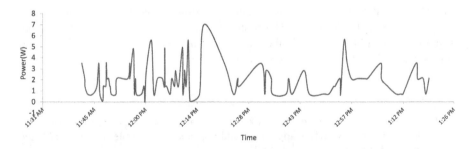

FIGURE 4.22   DL added/removed during DD to maintain bus voltage.

The DD has then been deployed to alleviate the voltage fluctuation in the microgrid. Load addition/removal during DD operation is depicted in Figure 4.22 and shows the record of power consumption on the microgrid; the consumption is matched to the real-time generation by connection and disconnection of DL as decided by the DD algorithm and implemented through LCU. The rising power in

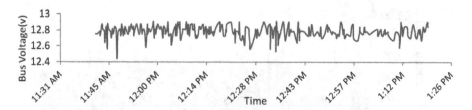

FIGURE 4.23   Bus voltage with DD on DC microgrid.

FIGURE 4.24   The rooftop m-WTG in DC microgrid.

Figure 4.22 depicts the addition of DL and the falling power depicts removal of DL. The profile of grid voltage with DD is shown in Figure 4.23; obviously, the voltage is maintained within the permissible range. The m-WTG on the rooftop that is used for the test and measurements is shown in Figure 4.24.

BIBLIOGRAPHY

1. Shariatzadeh, F., P. Mandal and A.K. Srivastava. "Demand response for sustainable energy systems: A review, application and implementation strategy." *Renewable and Sustainable Energy Reviews* 45 (2015): 343–350.
2. Samad, T. and A.M. Annaswamy. "Controls for smart grids: Architectures and applications." *Proceedings of the IEEE* 105(11) (2017): 2244–2261.
3. Qdr, Q. "Benefits of demand response in electricity markets and recommendations for achieving them." US Dept. Energy, Washington, DC, USA, Tech. Rep (2006).
4. Brooks, A., Lu, E., Reicher, D., Spirakis, C. and Weihl, B. "Demand dispatch." *Power and Energy Magazine, IEEE* 8(3) (2010): 20–29.
5. Gkatzikis, L. and I. Koutsopoulos, "The role of aggregators in smart grid demand." *IEEE Journal on Selected Areas in Communications* 31(7) (2013): 1247–1257.

6. Toth, P. "Dynamic programming algorithms for zero-one knapsack problem." *Computing* 25(1) (1980): 29–45.

7. Jia, M., A. Komeily, Y. Wang, and R.S. Srinivasan. "Adopting internet of things for the development of smart buildings: A review of enabling technologies and applications." *Automation in Construction* 101 (2018) (2019): 111–126.

8. Li, W. and X. Zhang. "Simulation of the smart grid communications: Challenges, techniques, and future trends." *Computers & Electrical Engineering* 40(1) (2014): 270–288.

9. Nithin, S., S.K. Kottayil and R. Lagerstöm."Direct load control on smart micro grid supported by wireless communication and real time computation." *Proceedings of the International Conference on Interdisciplinary Advances in Applied Computing—ICONIAAC*, 2014.

10. Nithin, S., K.K Sasi and T.N.P. Nambiar. "Development of a smart grid simulator." *Proceedings of National Conference on Power Distribution*, 2012, CPRI, India.

11. Phadke, A.G. "Synchronized phasor measurements in power systems." *IEEE Computer Applications in Power* 6(2) (1993): 10–15.

5 Dynamic Energy Management in Smart Microgrids

D. Prasanna Vadana

CONTENTS

5.1 INTRODUCTION

Power balancing in smart microgrids (SMG) when working in grid connected or islanded mode is essential for efficient operation. This requires dynamic control and the subsequent automated operation of SMG that facilitates the power balancing mechanism in SMG. One such strategy that assists in handling the energy imbalance that occurs due to variations in generation and load demand in both grid-connected and islanded SMGs is the dynamic energy management (DEM) scheme. A universally adopted method of dynamic power balancing on SMG is by use of energy storage devices and systems. The charge-discharge rates of energy storage systems vary widely among the commercially available systems. A DEM scheme has to involve therefore more than one type of energy storage device in the system. The DEM scheme proposed here uses two such storage systems – battery and pumped hydro. The development of the DEM system and its validation are presented in this chapter in detail.

5.2 DEM SCHEME

The DEM is defined as the control of charge-discharge transactions in the energy storage systems to oppose the frequency excursions on the grid in the real-time environment. The SMG under consideration is assumed to have generation schemes such as small hydro, solar PV, and wind. Two types of energy storage systems are connected to the microgrid – pumped hydro storage (PHS) and battery energy storage (BES). The hydro power plant is considered to have separate units for generation and storage. The microgrid chosen for implementing DEM is shown in Figure 5.1.

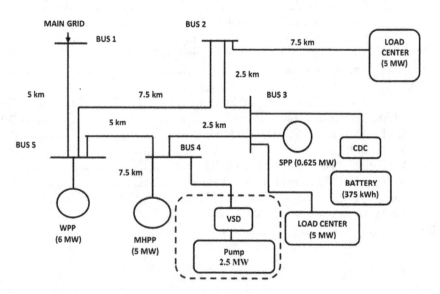

FIGURE 5.1   Microgrid chosen for the study. SPP: Solar power plant; MHPP: Micro hydel power plant; WPP: Wind power plant; VSD: Variable speed drive; CDC: Charge-discharge controller; PHS: Pumped hydro storage.

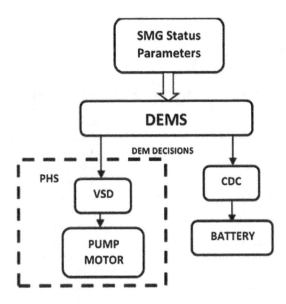

FIGURE 5.2   Dynamic energy management scheme.

Figure 5.2 shows the DEM scheme performed on the storage systems installed on a SMG to help maintain the supply-demand balance.

DEM scheme requires the status parameters of SMG to output a DEM decision which represents the action to be taken on the energy storage systems to balance the supply and demand. The DEM scheme is implemented on dynamic energy management system (DEMS) to perform the task of decision making.

The DEMS is a platform that receives the real-time data from the communication network present in SMG, forms SMG status parameters as listed in Table 5.1 and outputs a decision based on the control algorithm implemented in it. The real-time data required to form these status parameters include the following:

- Magnitude and phase angle of grid voltage
- Magnitude and phase angle of microgrid voltage
- Load current
- Current delivered to the PHS
- Battery charging/discharging current
- State of charge (SoC) of battery
- Frequency.

The DEMS receives these real-time data from data loggers installed on the SMG which are connected in the communication network. Voltages on the grid and microgrid decide the status parameter S_{W1}. The load current and pumped hydro current decide S_{W2} and S_{W3}, respectively. Charging/discharging current and SoC of the battery decide the battery status parameters S_{Ic} and S_{SOC}, respectively. The grid frequency decides S_f.

TABLE 5.1

SMG Status Parameters

Status Parameter	Definition	States
S_{W1}	Status of power exchange between the main grid and SMG	1: Power is imported from the main grid −1: Power is exported to the main grid 0: No power exchange
S_{W2}	Status of local demand	1: Power is drawn by local load 0: Power is not drawn by local load
S_{W3}	Status of PHS	1: Pumping 0: Idling
S_{Ic}	Status of battery	1: Charging −1: Discharging 0: Disconnected
S_{SOC}	State of charge of the battery	1: Fully charged 0: Partially charged −1: Below discharge limit
S_f	Status of frequency	1: Below nominal frequency −1: Above nominal frequency 0: Equal to nominal frequency

The status parameter is a qualitative indication of the state of SMG represented digitally as 1, −1 or 0. This scheme works with the status of the SMG rather than the actual values of the parameters, thereby making the decision in a qualitative manner. This permits DEMS to be structured based on the centralized energy management system (CEMS) architecture engraved with the features of decentralized CEMS (De-CEMS). This nature of the scheme makes DEMS a generalized system versatile to work with any SMG size and configuration. Also, the memory required for storing real-time measured voltage and current data would be greater than that needed for storing the status values. This paves the way for the base platform used to implement DEMS to be cost-efficient. Processing of status values also requires less time which results in faster operation of DEMS in predicting a decision.

The three operating modes of the SMG with all possible combinations of the main status parameters – namely, S_{W1}, S_{W2} and S_{W3} – are identified and listed in Table 5.2. The islanded mode depicts the operation of SMG when there is no power exchange with the grid. The grid import mode depicts the operation when power is imported from the grid as the local generation (LG) is insufficient to meet the local demand (LD). The grid export mode depicts the operation when power is exported to the grid as there is LG in excess of LD.

TABLE 5.2

SMG Operating Modes

Modes of Operation of SMG	Mode	Description & Observation	S_{W1}	S_{W2}	S_{W3}
SMG in islanded mode	1	LG is used to charge battery.	0	0	0
	2	LD is nil; LG is delivered to PHS.	0	0	1
	3	LD is met by LG.	0	1	0
	4	LG > LD	0	1	1
SMG in grid import mode	5	Grid power is imported to charge battery.	1	0	0
	6	Grid power is imported to charge PHS.	1	0	1
	7	LD > LG	1	1	0
	8	(LD + PHS) > LG	1	1	1
SMG in grid export mode	9	LG is exported to grid.	−1	0	0
	10	LG is exported to grid as well as stored in PHS.	−1	0	1
	11	(LG − LD) is exported to grid.	−1	1	0
	12	(LG − LD) is exported to grid and also stored in PHS.	−1	1	1

5.2.1 Generation of Data Patterns in Grid-Connected Mode

When the SMG is operated in grid-connected mode, it can be either in the grid import mode or grid export mode. The remaining three status parameters (i.e., S_{Ic}, S_{SOC}, S_f), each having 3 states (1, −1, and 0), can have the probable occurrences of $3^3 = 27$ as shown in Table 5.3. Finally, all the six status parameters (S_{W1}, S_{W2}, S_{W3}, S_{Ic}, S_{SOC} and S_f) are combined to form the data patterns for the grid-connected mode, comprising the Export and the Import modes.

The DEM decision for each status of the grid-connected SMG is made based upon the control operations of the energy storage systems and the related charge-discharge transactions as listed in Table 5.4. The current status of the grid is maintained if the decision is 1. This is required when the real-time grid frequency is the nominal frequency, that is 50 Hz. There are four control operations possible with the storage modules available in the microgrid system to perform DEM. Decisions 2, 3, 4 and 5 indicate the four control operations individually. Decision 2 suggests increasing the speed of the PHS motor to utilize the excess generation on the grid, when the battery is already charging. Decision 3 suggests decreasing the speed of the PHS motor when there is a need to reduce the rate at which the energy is stored.

Decision 4 suggests charging the battery. This is done when the status of frequency is "−1" (i.e., when grid frequency is above its nominal value). Decision 5

TABLE 5.3

Possible Combinations of S_f, S_{SOC}, S_{lc} (Grid-Connected Mode)

S. No.	S_f	S_{lc}	S_{SOC}
1.	0	0	0
2.	0	0	1
3.	0	0	−1
4.	0	1	0
5.	0	1	1
6.	0	1	−1
7.	0	−1	0
8.	0	−1	1
9.	0	−1	−1
10.	1	0	0
11.	1	0	1
12.	1	0	−1
13.	1	1	0
14.	1	1	1
15.	1	1	−1
16.	1	−1	0
17.	1	−1	1
18.	1	−1	−1
19.	−1	0	0
20.	−1	0	1
21.	−1	0	−1
22.	−1	1	0
23.	−1	1	1
24.	−1	1	−1
25.	−1	−1	0
26.	−1	−1	1
27.	−1	−1	−1

TABLE 5.4

DEM Decisions (Grid-Connected Mode)

DEM Decision	Description
1	Maintain status quo
2	Increase charging of PHS
3	Decrease charging of PHS
4	Charge the battery
5	Discharge the battery
6	Increase the speed of PHS and charge the battery
7	Decrease the speed of PHS and discharge the battery

TABLE 5.5

Sample Instances for Grid-Connected Mode

Case	S_{W1}	S_{W2}	S_{W3}	S_f	S_{lc}	S_{SOC}	DEM Decision
A	1	1	0	0	−1	0	1
B	−1	0	0	−1	0	1	2
C	1	1	1	1	0	0	3
D	−1	0	1	−1	−1	0	4
E	1	1	0	1	1	1	5
F	−1	1	0	−1	−1	−1	6
G	1	1	1	1	1	1	7

suggests discharging the battery, when the status of frequency is "1" (that is, when the frequency is below its nominal value). The requirement of the decisions 6 and 7 are explained along with examples as given in Table 5.5.

Case A: Grid is supplying power to the LD with the frequency at 50 Hz. The battery is discharging and PHS is disconnected. As the SoC of the battery can support further discharge, control action is to maintain the same status of the system.

Case B: The LG is exported to the grid when the frequency is above 50 Hz which is not desirable. As the battery is full, the control action is to increase the speed of PHS so that power export to the grid can be avoided.

Case C: Grid is supplying power to the LD and also charging PHS unit while frequency is below 50 Hz. The storage system should then not be charged using the grid power. Minimization of pumping power should be the control action so that power imported will eventually decrease.

Case D: Power is exported to the grid while the frequency is high; PHS is pumping, but the battery is discharging. Charging the battery is the control action as grid export should be prevented at higher frequency.

Case E: The frequency of the grid is below 50 Hz. The control action is to discharge the battery to share the local load and in turn to reduce the burden on the grid.

Case F: The grid frequency is above 50 Hz with battery in its discharging state and PHS in idle state. This is a favorable condition for charging the storage systems. Increasing the pumping power and charging the battery together should be the control action assigned for this combination of status parameters.

Case G: The grid needs help as its frequency is less than the ostensible value of 50 Hz. Reduction of pumping power along with discharge of the battery should be performed on the SMG to reduce the burden on the grid.

The combination of all the six status parameters results in a set of 199 data patterns with the appropriate decisions assigned; the impossible occurrences are ignored. Table 5.6 shows the details of the number of patterns distributed under each decision based on the strategy mentioned.

TABLE 5.6

Distribution of Data Patterns under Each Decision (Grid-Connected Mode)

Decision	No. of Patterns
1	21
2	47
3	30
4	25
5	19
6	36
7	21
Total	**199**

5.2.2 GENERATION OF DATA PATTERNS IN ISLANDED MODE

In islanded mode, DEM scheme is used when the frequency excursion is within a short band around the nominal grid frequency (say, between 49.7 and 50.1 Hz) as operated by a 50 Hz utility in the grid-connected mode. If the frequency is beyond these limits, DEM along with load management (LM) has to be performed. The DEM decisions are made as the control actions required to be taken on the islanded SMG to maintain the frequency. The frequency becomes a crucial parameter to control – as it is very sensitive to variations in both generation and demand – unlike the legacy grid. To have a finer account of frequency, the possible status values of frequency can have 5 ranges in this mode as listed in Table 5.7.

When the status of frequency is 1 or −1, it is evident that only fine tuning is required to balance the frequency. On the contrary, when the status of frequency is 2 or −2, it is apparent that the control action performed should bring a significant change in frequency. Adjusting the speed of PHS will not suffice in this case which also demands control of the local loads connected to the SMG. The LM becomes

TABLE 5.7

Status Values of S_f in Islanded Mode

Frequency (Hz)	S_f
>50.1	−2
$50 < f \leq 50.1$	−1
50	0
$49.7 \leq f < 50$	1
<49.7	2

TABLE 5.8

Possible Combinations of S_f, S_{SOC}, S_{lc} (Islanded Mode)

S.No.	S_f	S_{lc}	S_{SOC}	S.No.	S_f	S_{lc}	S_{SOC}
1.	2	0	0	24.	0	1	−1
2.	2	0	1	25.	0	−1	0
3.	2	0	−1	26.	0	−1	1
4.	2	1	0	27.	0	−1	−1
5.	2	1	1	28.	−1	0	0
6.	2	1	−1	29.	−1	0	1
7.	2	−1	0	30.	−1	0	−1
8.	2	−1	1	31.	−1	1	0
9.	2	−1	−1	32.	−1	1	1
10.	1	0	0	33.	−1	1	−1
11.	1	0	1	34.	−1	−1	0
12.	1	0	−1	35.	−1	−1	1
13.	1	1	0	36.	−1	−1	−1
14.	1	1	1	37.	−2	0	0
15.	1	1	−1	38.	−2	0	1
16.	1	−1	0	39.	−2	0	−1
17.	1	−1	1	40.	−2	1	0
18.	1	−1	−1	41.	−2	1	1
19.	0	0	0	42.	−2	1	−1
20.	0	0	1	43.	−2	−1	0
21.	0	0	−1	44.	−2	−1	1
22.	0	1	0	45.	−2	−1	−1
23.	0	1	1				

mandatory in the islanded mode as the hydro generation is limited and RE sources like solar and wind are intermittent. The operating modes for the three major status parameters (S_{W1}, S_{W2}, S_{W3}) for islanded mode are retained from Table 5.2. The possible variations of the remaining three parameters (S_f, S_{lc}, S_{SOC}) are shown in Table 5.8. On combining the six status parameters with all the possible combinations, the data patterns for islanded mode are generated. Every case of data set is assigned a decision based on the DEM principles as listed in Table 5.9.

Demand response (DR) is a modern method of load management and direct load control (DLC) is one of its techniques. A special DR program linking captive solar PV generation with DLC is implemented here. Rooftop solar PV plant installed on the SMG consumer's premises is used to charge the consumer's local battery storage. Yet, the captive solar plant is allowed to energize the consumer's load only upon receiving an instruction from DEMS. Under normal conditions of frequency, the consumer's load is energized by the SMG. When required, a signal is sent to the local

TABLE 5.9
DEM Decisions (Islanded Mode)

DEM Decision	Description
1	Maintain status quo
2	Charge the battery
3	Discharge the battery
4	Adjust the speed of PHS
5	Perform load management
6	Charge the battery and perform load management
7	Discharge the battery and adjust the speed of PHS

controller (LC) at the consumer's premises by DEMS either to connect/disconnect the SMG supply to the consumer load. The LC performs the following two actions:

1. Upon receiving the signal to switch the SMG supply OFF, the consumer load is energized by the local battery, which is charged by the local solar PV system.
2. Upon receiving the signal to switch the SMG supply ON, the consumer load is energized by the SMG, disconnecting it from the local battery.

By this solar linked DLC, the customer can enjoy the benefits of uninterrupted power supply even when disconnected from the SMG.

The DEM decision of maintaining the same state of the SMG (Decision 1) is required at instances when the frequency is at its nominal value. Decision 2 suggests charging of battery. This is required for the fine tuning of frequency in smaller steps especially when the status of frequency is "−1" provided PHS is already receiving power. Decision 3 suggests discharging of battery – this facilitates the fine tuning of frequency when it is just below 50 Hz. Decision 4 indicates the adjustment of speed of PHS. Decision of whether to increase or decrease the speed of PHS is determined using a secondary controller. Decision 5 suggests performing LM – this decision is made when the frequency deviation is large. Similarly, the LM module is also enabled with a secondary controller which decides whether the disconnected loads are to be reconnected or live loads are to be disconnected.

The decisions 6 and 7 are discussed in detail along with Table 5.10 which illustrates the smart control operations.

Case A: Local demand is met by local generation with the frequency at its nominal value. The decision should be to maintain the same state of SMG.

Case B: Local generation is in excess of the local demand indicated by the status of frequency. The control action suggested is to charge the battery as the frequency is within the range of 50–50.1 Hz. This will facilitate a smaller step change in frequency.

Case C: Local demand is in excess of local generation indicated by the status of frequency with a value of "1" (between 49.7 and 50 Hz). Status of battery

TABLE 5.10

Sample Instances for Islanded Mode

Case	S_{W1}	S_{W2}	S_{W3}	S_f	S_{lc}	S_{SOC}	DEM Decision
a	0	1	0	0	0	1	1
b	0	1	0	-1	0	-1	2
c	0	1	0	1	1	0	3
d	0	1	1	2	-1	1	4
e	0	0	1	-2	0	1	5
f	0	1	0	-2	-1	-1	6
g	0	1	1	2	1	1	7

indicates charging. Decision suggested is to discharge the battery which results in a step increase in frequency.

Case D: Local demand together with PHS is in excess of local generation which is aided by the discharging battery. The frequency is less than 49.7 Hz which needs an action leading to a significant change. Either LM (decision 5) or adjusting the speed of PHS (decision 4) can be performed. Priority to stay connected with the SMG is given to the local loads. This results in a decision to adjust the speed of PHS.

Case E: The presence of local generation is confirmed through the status of PHS. All the loads are disconnected from the SMG. The status of frequency is "−2" indicating the space to reconnect the local loads to the SMG. No action can be taken with the battery as it is already fully charged.

Case F: As S_{W2} is a qualitative parameter, only the presence or absence of local loads is indicated by its status. There is a possibility that only some loads are connected, and the rest are disconnected. In this instance, the frequency is greater than the nominal value with battery being discharged. Decision suggested is to charge the battery and to perform LM. This decision gives a scope for the local loads to be connected to the SMG if they were disconnected previously. Though PHS is switched off, priority to get connected to the SMG is given to the local loads.

Case G: The frequency of SMG is below the permissible limit with the presence of local loads and charging of storage systems. Burdening the SMG with charging the storage modules under low frequency conditions needs to be avoided. The decision is to discharge the battery and to adjust the speed of PHS simultaneously which produces a significant change in the frequency.

A total of seven DEM decisions are identified and are assigned to the data occurrences. The strategy followed in assigning the DEM decisions to the possible occurrences of the SMG is summarized as follows:

- Priority to stay connected/to get reconnected with the SMG is given to the local loads rather than the energy storage systems.

TABLE 5.11

Distribution of Data Patterns under Each Decision (Islanded Mode)

Decision	No. of Patterns
1	27
2	23
3	21
4	15
5	45
6	17
7	11
Total	**159**

- When $S_f = 1$ or -1, DEM with respect to battery storage is suggested to fine tune the frequency with small step change.
- When $S_f = 2$ or -2, DEM along with LM is suggested to coarse tune the frequency with large step change.

The combination of all the six status parameters results in a set of 159 data patterns with the appropriate decisions assigned. The impossible occurrences are ignored. Table 5.11 shows the details of the number of patterns distributed under each decision based on the strategy mentioned.

5.3 DEMS ALGORITHM

5.3.1 DEMS Algorithm for Grid-Connected Mode

Decision making should be automated in such a manner that whenever the six status parameters are formed, an appropriate control action should be suggested by a suitable control algorithm implemented in the DEMS. This can be done by any one of the following two methods: (i) Artificial neural networks (ANN) and (ii) support vector machines (SVM). The realization of the DEM strategy using the two methods is discussed in detail and the appropriate method is chosen based on accuracy.

5.3.1.1 Artificial Neural Network

This application comes under the class of supervised learning as the target vectors are known for the specific input patterns. The algorithm chosen to implement the mapping between the SMG status and the DEM decision is back propagation (BP) algorithm. The entire set of 199 patterns obtained for grid-connected mode operation is divided into two groups: One group of 159 patterns for training and another group of 40 patterns for testing, both with equal distribution of all the decisions. Based on this division of data patterns, two different neural network models are designed and trained.

Both the models have 6 input units to intake the six status parameters and 3 output neurons to output the decision which is interpreted in a bipolar manner as shown

TABLE 5.12

Bipolar Interpretation of DEM Decisions

Decision	Y_1	Y_2	Y_3
1	−1	−1	1
2	−1	1	−1
3	−1	1	1
4	1	−1	−1
5	1	−1	1
6	1	1	−1
7	1	1	1

in Table 5.12. The number of hidden layers and the number of neurons in each hidden layer is determined using trial and error methods. The architecture of Model-1 (6-30-3) consists of 6 neurons in the input layer, 30 neurons in the hidden layer and 3 neurons in the output layer as shown in Figure 5.3.

The architecture of Model-2 consists of 3 sub-models each having 6 neurons in the input layer and 1 neuron in the output layer. The architectures of sub-models 2.1, 2.2, and 2.3 have 7, 6 and 5 neurons, respectively in the hidden layers as shown in Figure 5.4a–c. Every circle in the architectures shown in Figures 5.3 and 5.4 represents an artificial neuron. For this application, the activation function used for both hidden layer and output layer neurons in both models is a hyperbolic (or bipolar) sigmoid function whose output ranges from −1 to +1 (similar to the decisions listed in Table 5.12) as shown in Figure 5.5. The mathematical expression for the bipolar sigmoid activation function is:

$$f(x) = \frac{1 - \exp(-\sigma x)}{1 + \exp(-\sigma x)} \tag{5.1}$$

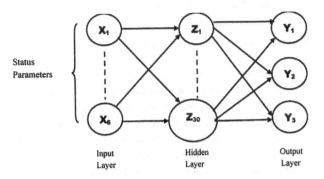

FIGURE 5.3 Architecture for Model-1 with 6 input neurons, 30 hidden neurons, and 3 output neurons.

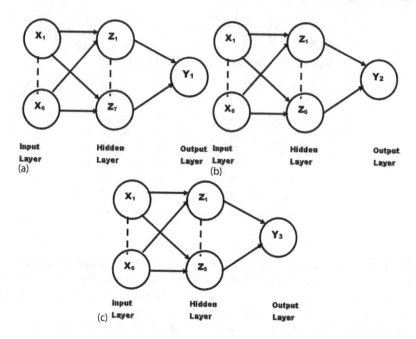

FIGURE 5.4 (a–c) Architectures of sub-models 2.1, 2.2 and 2.3 each having 7, 6 and 5 neurons in the hidden layer, respectively.

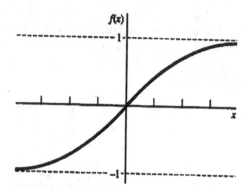

FIGURE 5.5 Bipolar sigmoid activation function.

After training, application of the net involves only the computation of the feed forward phase. Even if training is slow, a trained network can produce its output very rapidly. Model-2 which contains three sub-models is chosen to be the best model as it has the highest accuracy of classification when tested with various combinations of unknown input patterns. It is evident from Table 5.13 that Model-2 is unable to classify only 8 patterns out of 40 unknown patterns.

The mathematical basis for BPN is the gradient descent optimization algorithm. The gradient of the function – in this case the function is the error and the variables are the weights in the network – gives the direction in which the function increases

TABLE 5.13

Simulation Results for DEMS Realization Using ANN

Model	Testing Accuracy (%)	
	Untrained Patterns	Entire Dataset
Model-1 (6-30-3)	65(26/40)	84.11(115/199)
Model-2 (6-7-1, 6-6-1, 6-5-1)	80(32/40)	90.1(166/199)

very rapidly; the negative of the gradient gives the direction in which the function decreases very rapidly. The BPN suffers from getting trapped in multiple local minima affecting the accuracy of DEMS, a critical issue for this SMG application. Also, if the dimensionality of the input vector changes, the training becomes complex in ANNs. The requirement is that DEM should be realized using an algorithm which gives 100% accuracy in suggesting the controlling actions based on the status of SMG in a more intelligent way. The suitability of SVM to realize the DEM scheme is studied next.

5.3.1.2 Support Vector Machines

The SVMs are a useful technique for classification and regression as the roots are from statistical learning theory and optimization techniques, unlike ANNs which are heuristic. The task of performing DEM (i.e., to take the appropriate decision that manages the frequency imbalance) is a multi-class classification problem. A classification task usually involves separating data into training and testing sets. Each instance in the training set contains one "target value" (i.e., the class labels) and several "attributes" (i.e., the features or observed variables). The goal of SVM is to produce a model (based on the training data) which predicts the target values of the test data, given only the test data attributes.

This is performed by the creation of decision planes that define decision boundaries. A decision plane is one that separates classes in a set of objects having different class memberships. The SVM always tries to achieve a linear classification between the classes. Most of the real-time applications can be classified only in a non-linear manner. Kernel functions are employed by SVM in such cases to convert the non-linear classifier to a linear classifier. The responsibility of kernel functions is to increase the dimensionality of the non-linear input space which is termed as the "feature space." The non-linear input can be linearly classified in the feature space as illustrated in Figure 5.6.

The problem of deciding the appropriate action to be performed becomes a non-linear classification which can be linearly classified using kernel functions. Some of the basic kernel functions are as follows:

1. Linear: $k(x_i, x_j) = x_i^T x_j$ (5.2)
2. Polynomial: $k(x_i, x_j) = (\gamma x_i^T x_j + r)^d; \gamma > 0$ (5.3)
3. Radial basis function (RBF): $k(x_i, x_j) = e^{\left(-\gamma \|x_i - x_j\|\right)^2}, \gamma > 0$ (5.4)

Here, γ and d are kernel parameters.

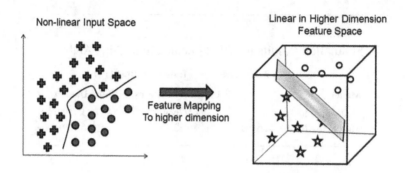

FIGURE 5.6   Non-linear to linear transformation using SVM methodology.

In general, the RBF kernel represented in Equation 5.4 is a reasonable first choice for realizing the DEM scheme. This kernel non-linearly maps samples onto a higher dimensional space. Hence, it can handle the case when the relation between class labels and attributes is non-linear, unlike the linear kernel. There are two parameters for an RBF kernel: Cost or penalty parameter (C) and kernel function parameter (γ). The variable C is a user-defined control parameter that specifies a trade-off between the hyperplane violations and the size of the margin. Increasing the value of C increases the cost of misclassifying points and forces the creation of a more accurate model that may not generalize well. It is not known beforehand which C and γ are the best values for a given problem; consequently, some kind of model selection (parameter search) must be done.

The goal is to identify good C and γ so that the classifier can accurately predict unknown data (i.e., testing data). A common strategy is to separate the data set into parts, of which some are considered unknown. The prediction accuracy obtained from the unknown set more precisely reflects the performance on classifying an independent data set. An improved version of this procedure is known as cross-validation. In v-fold cross-validation, the training set is divided into v subsets of equal size. Sequentially one subset is tested using the classifier trained on the remaining ($v-1$) subsets. Thus, each instance of the whole training set is predicted once and the cross-validation accuracy is the percentage of data which are correctly classified.

In this application, the entire data set of 199 patterns is divided into 5 subsets of equal size such that each subset contains equal distribution of the data from all the 7 decisions as shown in Table 5.14. Five optimized SVM models are developed by combining four data groups at a time and each model is tested with the fifth group constituting the unknown data patterns as shown in Figure 5.7. Every model is named as per the training and testing patterns as: ABCD_E which implies, this model is trained using the patterns from the groups A, B, C and D, and tested using the unknown patterns from the group E. The software tool used to create the SVM

TABLE 5.14

Distribution of Data Patterns into Groups (Grid-Connected Mode)

Group/Decision	1	2	3	4	5	6	7	Total
A	5	9	6	5	3	7	4	39
B	4	10	6	5	4	7	4	40
C	4	9	6	5	4	7	5	40
D	4	9	6	5	4	8	4	40
E	4	10	6	5	4	7	4	40
Total	21	47	30	25	19	36	21	199

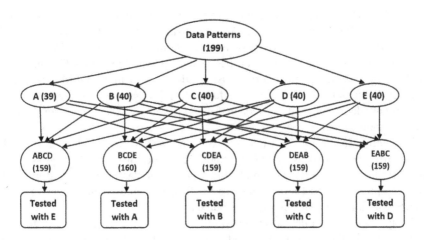

FIGURE 5.7 Distribution of data patterns in grid-connected mode.

model is LIBSVM-MAT-2.9.1 integrated with MATLAB®. Optimization is achieved by selecting the appropriate RBF kernel parameters C and γ chosen using grid-search method in this application.

The simulation results for all the models obtained during training and testing are listed in Table 5.15. The simulation results show the cross-validation accuracy and the testing accuracy of each model. During training, 5-fold cross validation is performed so as to increase the accuracy of each model. Accuracy being a critical issue in DEM implementation, the SVM model EABC_D is the right choice as this produces 100% accuracy. This implies that the model EABC_D is capable of identifying all the unknown patterns and is classifying the data patterns perfectly when compared with the ANN realization.

TABLE 5.15

Simulation Results of SVM in Grid-Connected Mode

Model	γ (Gamma)	C (Cost)	Cross Validation Accuracy (%)	Testing Accuracy of Unknown Patterns (%)	
ABCD_E	1	2	91.8239	80 (32/40)	8
BCDE_A	0.0625	64	95.625	94.8718 (37/39)	2
CDEA_B	1	8	86.1635	80 (32/40)	8
DEAB_C	1	16	89.3082	85 (34/40)	6
EABC_D	0.0625	64	93.0818	100 (40/40)	0

5.3.2 DEMS ALGORITHM FOR ISLANDED MODE

Islanded mode of SMG is the result of intentional or unintentional grid disconnection of the SMG. Status parameter S_{W1} is considered to be "0" in islanded mode as there is no power exchange between the SMG and the main grid. Here, LM has to be performed on the DR and non-DR loads connected to the SMG. Automation of decision-making is mandatory in islanded mode to make DEMS handle the frequency imbalance, being sensitive to load and generation changes. Being the appropriate algorithm in grid-connected mode, SVM is chosen to realize DEMS in islanded mode, also.

The entire set of 159 data patterns as mentioned in Section 5.2.1 are divided into 5 different groups as shown in Table 5.16 for performing 5-fold cross validation. Similar to grid-connected mode, the accuracy of classification is estimated for the five optimized SVM models formed by dividing the 127 data patterns into five groups with all decisions equally distributed amongst them.

Every model is named after the training data and testing data as shown in Figure 5.8. Similar to grid-connected mode, RBF kernel with parameters chosen using grid search method represented in Equation 5.4, is used for classification, as this kernel has the ability to map the input space of finite dimensions to a feature space of infinite dimensions where the classification becomes linear. The simulation

TABLE 5.16

Distribution of Data Patterns into Groups (Islanded Mode)

Group/Decision	1	2	3	4	5	6	7	Total
A	5	5	4	3	9	4	2	32
B	5	5	4	3	9	3	3	32
C	5	5	5	3	9	4	2	33
D	6	4	4	3	9	3	2	31
E	6	4	4	3	9	3	2	31
Total	27	23	21	15	45	17	11	159

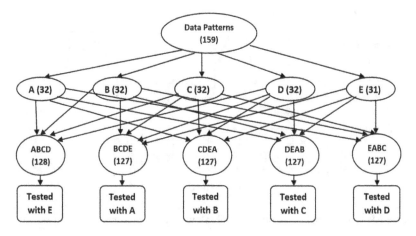

FIGURE 5.8   Distribution of data patterns in islanded mode.

TABLE 5.17
Simulation Results of SVM in Islanded Mode

Model	γ (Gamma)	C (Cost)	Cross Validation Accuracy (%)	Testing Accuracy of Unknown Patterns (%)	
ABCD_E	0.8	40	97.6563	77.4194 (24/31)	7
BCDE_A	0.5	4	96.85	65.625 (21/32)	11
CDEA_B	0.25	4	96.06	84.375 (27/32)	5
DEAB_C	0.15	32	96.85	93.75 (30/32)	2
EABC_D	0.5	4	97.6378	87.5 (28/32)	4

results of testing with the unknown patterns are presented in Table 5.17. It is evident that the model DEAB_C is the suitable model for implementing DEMS in islanded mode with the classification accuracy of 93.75%.

5.4   MODELING OF DEMAND SIDE RESOURCES

5.4.1   LOCAL DEMAND

The SMG is to meet a local demand of 10 MW which is grouped as

1. Domestic and commercial loads of 6 MW, and
2. Industrial loads of 4 MW.

Typical domestic and industrial load profiles as shown in Figures 5.9 and 5.10, respectively, are adopted to form the demand data for the simulation in the grid-connected mode.

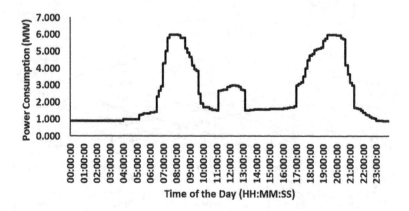

FIGURE 5.9 Daily load curve on the SMG (Domestic and commercial loads).

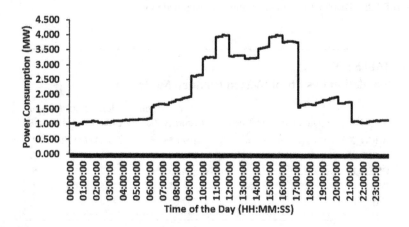

FIGURE 5.10 Daily load curve on SMG (Industrial load).

Whereas the islanded mode of operation stipulates the segregation of load demand into groups unlike in grid-connected mode, to distinguish between DR consumers and non-DR consumers in order to perform LM (as discussed in Section 5.2.2) as shown in Figure 5.11. The set of consumers connected to the SMG is divided into four categories, namely:

- *Domestic* loads representing the residential sector.
- *Essential* loads representing the hospitals, schools, etc.
- *Commercial* loads representing the theatres, shopping malls, etc.
- *Industrial* loads representing the mills, factories, etc.

The commercial loads of 1 MW and domestic loads of 0.4 MW are considered as DR members. Larger or smaller step changes in the frequency are accomplished by further subdividing the loads as shown in Table 5.18.

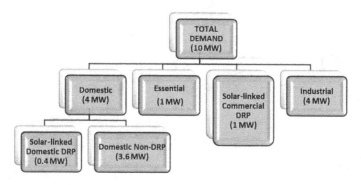

FIGURE 5.11 Segregation of local demand into DRP and non-DRP loads.

TABLE 5.18
Details of Segregation of Local Loads

Category of Load	No. of Groups of Load	Power Consumption by Each Group of Load (MW)	Total Power Consumption of the Category (MW)
DR commercial load	10	0.10	1.0
DR domestic load	8	0.05	0.4
Non-DR domestic load	10	0.36	3.6
Non-DR industrial load	10	0.40	4.0
Essential load	1	1.00	1.0
Total (in MW)			10

The DR and non-DR domestic load profiles are generated from the domestic load profile as shown in Figure 5.9. The commercial load profile is considered to be as shown in Figure 5.12. Essential load is exempted from inclusion in DR as it has to be served all the time and its profile is duplicated as that of the commercial load.

If all the DR consumers are disconnected from the SMG, the remaining loads are also due to be disconnected from the SMG, following a typical order of preference in times of adverse frequency conditions. Non-DR domestic consumers are given the top priority to get disconnected from the SMG followed by the industrial consumers. Once the non-DR consumers are disconnected from the SMG, they are allowed to reconnect with the SMG only after a stipulated time period. When only the essential loads remain connected to the SMG, its shedding also will be due in the case of a frequency collapse at that condition.

Vice-versa, the loads are included in the SMG, when frequency is on its higher side. Essential loads are given top priority to get connected to the SMG followed by the non-DR consumers and then the DR consumers. A secondary controller can be developed to handle the inclusion and exclusion of loads in such a manner that the DR consumers will have uninterrupted power supply and fiscal benefits.

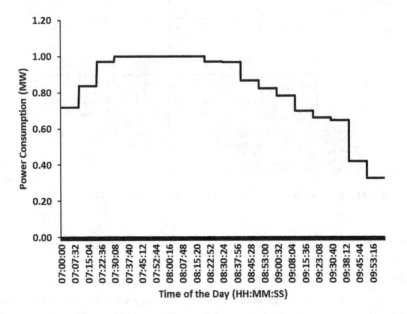

FIGURE 5.12 Daily load curve on SMG (Commercial load).

5.4.2 GENERATION

5.4.2.1 Wind Power Plant (WPP)

The power curve of a 2 MW wind turbine generator (WTG), modeled as Equation 5.5, is used to determine the respective power output of every measured wind speed at an interval of 10 min resulting in the WPP generation data for each day.

$$P = 0.001984v^5 - 0.008126v^4 + 0.1204v^3 - 0.7845v^2 + 2.407v - 2.79 \quad (5.5)$$

where:

  v: Wind speed, m/s, and
  P: Power output of WTG, W.

The inertial response of the WTG plays an important role in deciding the power output at any instant for a scenario of changing wind speed. The dynamics of the WTG is governed by Equation 5.6 in which the effect of friction is neglected.

$$J\omega\frac{d\omega}{dt} = P_2 - P_1 \quad (5.6)$$

where:

  J: Moment of inertia of the WTG, kg m^2,
  ω: Angular speed of rotation of the generator shaft, rad/s,
  $\frac{d\omega}{dt}$: Rate of change of angular speed, rad/s^2,
  P_2: WPP power output at time t_2, MW, and
  P_1: WPP power output at time t_1, MW.

Here t_1 represents the time instant ahead of t_2. The time interval from t_1 to t_2 is 4 s; it is subdivided into n divisions of 1 s interval as $t_{11}, t_{12},... t_{1n}$, leading to a time step of $\Delta t = 1\text{s}$. The power output at time t_1 increases in steps $(P_{11}, P_{12}, ... P_{1n})$ at various time instances, in accordance with the change in angular speed $(\Delta\omega)$, to become the power output P_2 at time t_2. For every change in wind speed, the corresponding change in power output should be known to determine the inertial response of the machine. This is determined by superimposing the power curve of the WTG with the speed-torque characteristics of the electric generator. The plot of angular speed against the power output of the generator is further developed; thus real-time power output for every change in wind speed is obtained. The generator being an induction machine has linear slip-torque characteristics from no load to rated conditions. The relation between angular speed and power output is modeled as Equation 5.7 and employed to determine the inertial response of the machine when there is a change in the wind speed.

$$\omega = 4.12 \times 10^{-5} P + 79.79 \tag{5.7}$$

The operation of WPP requires the real-time power output P_{2new} in response to wind speed changes. The algorithm used for computation of P_2 based on machine equations is shown in Figure 5.13.

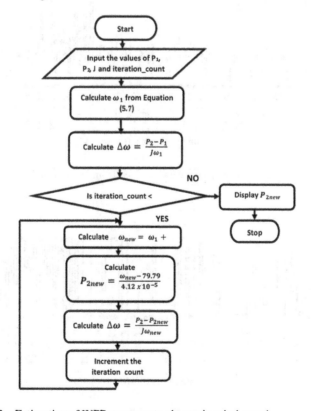

FIGURE 5.13　Estimation of WPP response to change in wind speed.

Additional symbols used in the flowchart are described as follows.

ω_1: Angular speed at time instant t_1, rad/s,
ω_2: Angular speed at time instant t_2, rad/s,
P_{2new}: Updated power output at the end of an incremented step, MW, and
ω_{new}: Updated angular speed at the end of an incremented step, rad/s.

Power output of 6 MW WPP corresponding to the time series wind speed data shown in Figure 5.14 after considering the effect of inertia is shown in Figure 5.15.

To demonstrate the method, an instance of change in wind speed from v_1 (8.4 m/s) to v_2 (8.0 m/s) shown as a circled region in Figure 5.15 is considered. The WPP is considered to operate at steady state at v_1 generating a power output of P_1 (3.829 MW) as per the power curve. It is then considered that v_1 decreases to v_2. The steady state power output of WPP at v_2 is P_2 (3.360 MW) as per the power curve. The response time of the WPP to change its output from P_1 to P_2 has been estimated using the developed algorithm and the result is shown in Figure 5.16.

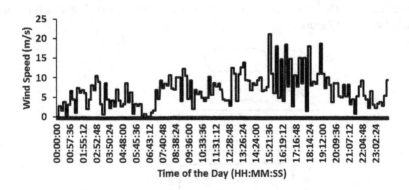

FIGURE 5.14 Time series wind speed data obtained from weather monitoring station.

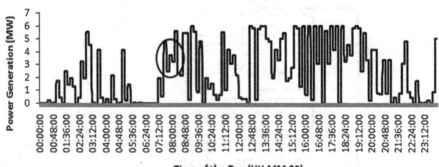

FIGURE 5.15 Instantaneous power output of WPP (Dynamic response).

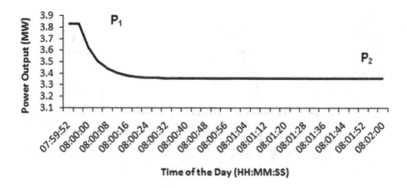

FIGURE 5.16 WPP response to wind speed change.

5.4.2.2 Solar Power Plant (SPP)

Solar PV module PVL-68 from Uni-Solar PowerBond™ of 68 W_p under standard test conditions is used for the modeling of SPP. The estimated power output for 0.625 MW_p SPP is shown in Figure 5.17. The reduction in power generation due to rise in cell temperature has been ignored in the estimate.

5.4.2.3 Micro Hydro Power Plant (MHPP)

The micro hydroelectric power station is scheduled to have a generation as shown in Figure 5.18 for grid-connected mode and Figure 5.19 for islanded mode, respectively, to meet the demand. The MHPP operates at its maximum capacity as shown in Figure 5.19 in the early hours of the day as generation from WPP and SPP are insufficient to serve the local demand. This is based on the assumption that the total generation from wind, solar, and hydro will be able to meet the demand with some reserve.

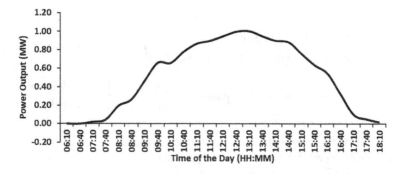

FIGURE 5.17 Generation profile of SPP of 0.625 MW_p.

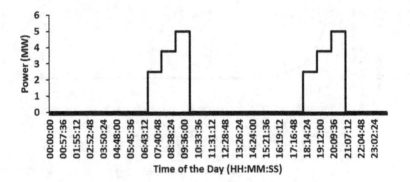

FIGURE 5.18 Generation schedule for MHPP (Grid-connected mode).

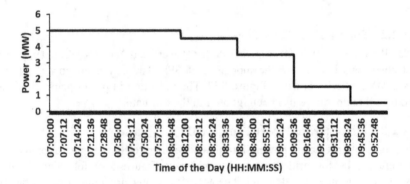

FIGURE 5.19 Generation schedule for MHPP (Islanded mode).

Excess and deficit power on the SMG (described by Equation 5.8) are adjusted with the legacy grid in grid-connected mode, whereas the total supply on the SMG is dynamically matched with the local demand in islanded mode.

$$\begin{Bmatrix} \text{Excess or} \\ \text{deficit power} \end{Bmatrix} = \begin{Bmatrix} \text{MHPP} \\ \text{generation} \end{Bmatrix} + \begin{Bmatrix} \text{SPP} \\ \text{generation} \end{Bmatrix} + \{\text{WPP generation}\}$$

$$\pm \{\text{Battery power}\} - \begin{Bmatrix} \text{Local} \\ \text{demand} \end{Bmatrix} - \begin{Bmatrix} \text{Power delivered} \\ \text{to PHS} \end{Bmatrix} \tag{5.8}$$

+ve battery power indicates that battery is discharging.
−ve battery power indicates that battery is charging.
+ve value on R.H.S. of (5.8) indicates excess generation.
−ve value on R.H.S. of (5.8) indicates generation deficit.

5.4.3 Energy Storage

5.4.3.1 Battery Energy Storage

Sodium sulphur (NaS) battery is chosen for energy storage considering its ability to with stand high fluctuations in charging and discharging conditions there by enabling smooth penetration of RE onto the grid. The equivalent circuit for the NaS battery is shown in Figure 5.20.

The selected model takes into account the non-linear battery element characteristics during charging and discharging as well as the internal resistance which depends on the temperature changes and SoC of the battery. The purpose of the diodes is to make NaS battery possess different values of internal resistances during charging and discharging operations. R_{lc} is the deterioration resistance based on the number of charge-discharge cycles; it is neglected for shorter periods.

An initial investigation is carried out with generation schedule and load curve in order to determine the size of energy storage to be in the range of 300–400 kWh. So, 320 cells of NaS totaling to a capacity of 375 kWh which charges or discharges at a constant power of 1 kW against instructions from DEMS is considered. Look up tables (LUTs) for E, R_C and R_D are obtained from the battery characteristics. E_{new}, the energy put into the battery in 4 s (as DEMS action occurs at every 4th s) is,

$$E_{new} = E_{old} + \frac{4VI_c}{3600} \tag{5.9}$$

where E_{old} is the previous value of E,

$$I_c = \frac{V - 320E}{320R_C} \tag{5.10}$$

is the battery charging current, and

$$V = \frac{320E + \sqrt{(320E)^2 + 4(320)(1000)R_C}}{2} \tag{5.11}$$

is the voltage applied across the battery terminals to charge it at the rate of 1 kWh per hour.

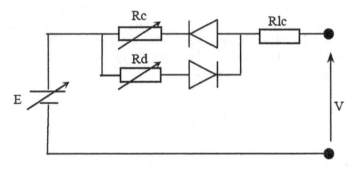

FIGURE 5.20 Selected NaS battery model E: Cell electro motive force, V; R_C: Charging resistance, Ω; R_D: Discharging resistance, Ω; V: Terminal voltage, V; R_{lc}: Life cycle resistance, Ω.

The SoC of the battery is computed as

$$\text{SoC} = \frac{(E_{\text{new}} \times 10^{-3})}{375} \qquad (5.12)$$

For the discharge case,

$$V = \frac{320E + \sqrt{(320E)^2 - 4(320)(1000)R_D}}{2} \qquad (5.13)$$

and

$$I_c = \frac{E(320) - V}{R_D(320)} \qquad (5.14)$$

5.4.3.2 Pumped Hydro Storage

The hydraulic turbine in a conventional pumped hydro station normally works as a pump when water is to be pumped to the upper reservoir and as a generator when power has to be generated. But such scheduled pumping and generation is impossible in this application as the schedule and the pumping rate are to be dynamically decided by the DEMS. The pump and the generator should be physically separate and are required to be operated in parallel. A variable speed drive is selected to operate the pump to control the power absorbed by PHS. This is modeled as 10 units of 250 kW induction motor running in parallel to pump the water.

The total capacity of PHS is chosen as 50% of that of the MHPP. The VSD operates based on a LUT containing the values of voltage (V) and frequency (f). The VSD uses the method of V/f control on the motor where V/f is kept constant while V and f are considered to vary the speed. In grid-connected mode, a single step change of V and f in the LUT (i.e., the adjacent higher or lower values) is enacted in order to increase or decrease the speed of the pump. This single step change would not suffice in islanded mode, as the PHS has to operate in synchronism with the variation in frequency. Therefore, a secondary controller is to be used to decide the number of steps of incrementing or decrementing the pumping rate when the DEMS instructs to adjust the speed of PHS. The operation of PHS aided by the LUT is carried out at a resolution of 0.2% (5 kW) of the total output delivery. For a motor power output of P_{op}, the water flow rate is computed as,

$$Q = \frac{1000 P_{op}}{\rho g h}, \text{m}^3/\text{s} \qquad (5.15)$$

where:
  h: Head (m),
  ρ: Density of water, 1000 kg/m^3, and
  g: Acceleration due to gravity, 9.8 m/s^2.

5.5 DEMS PERFORMANCE ON MICROGRID

5.5.1 SMG in Grid-Connected Mode

The processes involved in the implementation of DEMS in grid-connected mode are (i) *Data Input:* The necessary data of SPP, WPP and MHPP generation, local demand and frequency are provided; (ii) *Decision making:* WPP generation with the inertial response, total generation and total demand are computed. The status parameters are then computed to make the DEM decision; (iii) *Action handling:* The corresponding actions for the DEM decisions are handled in the respective energy storage modules and the status parameters are updated. Grid frequency measurement is monitored over a period of a day with a sampling period of 4 s as shown in Figure 5.21.

The corrective actions in the SMG are made by DEMS through its decisions on the battery and the PHS whenever there are deviations in frequency as shown in Figure 5.21 from its nominal value as apparent in Figure 5.22a and b, respectively. When several such microgrids are tied to the main grid, DEM performed in every microgrid contributes to frequency control of the entire power system network.

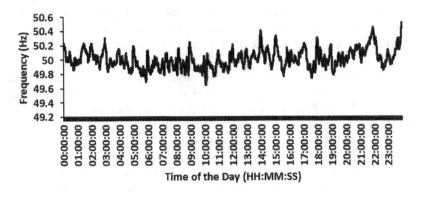

FIGURE 5.21   Frequency measurement over a day.

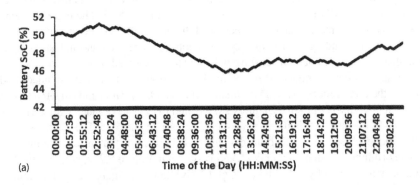

(a)

FIGURE 5.22   Effect of DEMS decisions on energy storage systems: (a) SoC of battery.
(*Continued*)

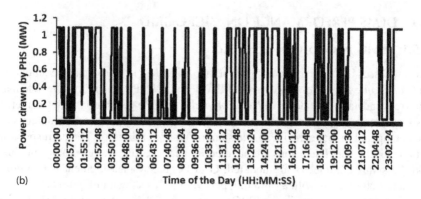

(b)

FIGURE 5.22 (Continued)   Effect of DEMS decisions on energy storage systems: (b) power drawn by PHS.

5.5.2   SMG in Islanded Mode

Islanded mode of SMG is a critical and challenging mode of operation. The frequency in grid-connected mode is decided together by all the entities connected to the grid and it is less influenced by the local activities of a microgrid. On the other hand, frequency in islanded mode needs to be computed in real time based on the instantaneous generation and load demand, in a simulation study.

The MHPP on the microgrid is a synchronous generator which generates power as scheduled. When there is a change in demand on this generator, owing to change in load or change in WTG and PV generation, the corresponding change in frequency is estimated using a simplified model that uses the equivalent circuit of the synchronous generator. This model uses fundamental network laws and converges to an appropriate frequency through an iterative process. As the generation is scheduled, the MHPP delivers constant power during the stipulated time period. When there is an increase in the demand, the generator tries to meet the demand with the stored kinetic energy (for a short duration) which in turn reduces the speed and causes reduction in the frequency. This reflects in the power angle of the generator. After a transient period, the machine gets settled to a steady state condition at a new frequency at which power equilibrium is attained. Frequency response of the SMG in islanded mode is shown in Figure 5.23. There are frequency excursions beyond the nominal frequency band. It needs to be ensured that the loads are capable of withstanding such frequency excursions. Commercial and industrial equipment that are highly frequency sensitive will need to be operated on UPS.

Similar to the grid-connected mode, implementation of DEM in an islanded mode also has the three processes: (i) *Data Input*, (ii) *Decision making*, and (iii) *Action handling*. Energy transactions in the battery storage and PHS in accordance with the DEMS decisions in the islanded SMG are shown in Figures 5.24 and 5.25. The sudden glitches in the frequency are due to the transient change in demand at that instant of time.

The original load demand on the SMG is shown in Figure 5.26a, whereas the load demand actually met by the SMG through LM is shown in Figure 5.26b. The demand unmet by the SMG is served from the captive solar plant at the consumer's premises through the solar-linked DR scheme.

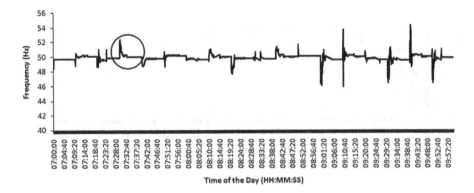

FIGURE 5.23 Real-time frequency excursions in islanded mode.

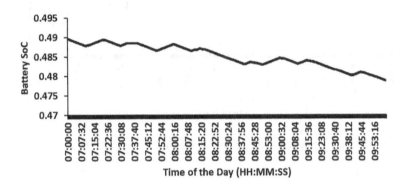

FIGURE 5.24 Effect of DEMS decisions on battery storage in islanded mode.

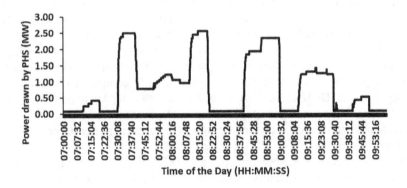

FIGURE 5.25 Effect of DEMS decisions on PHS in islanded mode.

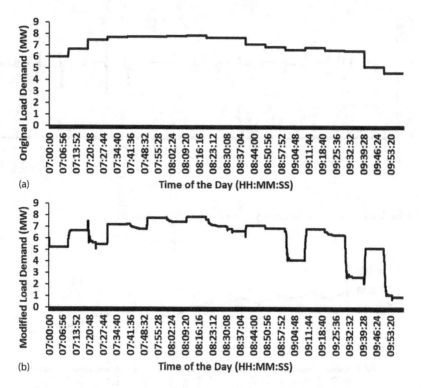

FIGURE 5.26   (a) Actual load demand and (b) load met by SMG through LM.

5.6   VALIDATION OF DEM ON EMULATED SMG

Validation of the DEM scheme carried out on SMGS, the laboratory scale realization of a microgrid presented in Section 3.6 of and shown in Figure 4.12, is presented in this section. The microgrid of Figure 5.1 is emulated on SMGS for the tests carried out.

Every bus in the SMGS is provided with one or two units of RTDCU, which is described in section 4.6.3 and shown in Figure 4.19. It is capable of measuring the magnitude and phase of the instantaneous voltage and the instantaneous current with time stamping and then using those measurements to compute the real-time frequency. Every RTDCU houses a dsPIC microcontroller enabled to perform simultaneous sampling of four different signals, essential for the RTDCU synchronized data acquisition. Also, a Zigbee communication module is connected to each RTDCU that transmits the measured data to an appropriate receiver.

A specific format is followed by all the RTDCUs to transmit the measured real-time data as shown in Equation 5.16. Here "P1" indicates that this data is from RTDCU 1 and "*" is the delimiter. The remaining quantities represent the magnitude and phase angle of the voltage and the current which are measured.

$$P1*225.06*-1.44*1.56*-0.24* \rightarrow \text{RTDCU No.} * |V|* \angle V * |I|* \angle I \qquad (5.16)$$

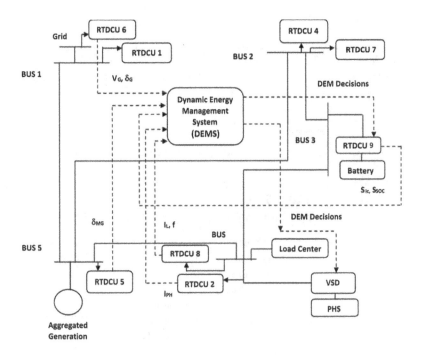

FIGURE 5.27 DEMS enabled SMG.

The SMGS is designed in such a manner that the RTDCUs transmit the measured real-time data in the specified format only at an interval of 4 s. The phase angles are measured with respect to a reference signal of 50 Hz frequency. Figure 5.27 shows the schematic of the DEMS enabled SMGS.

The DEMS, being an intelligent and energy aware system, should be implemented on a platform where configurability and response time are given utmost importance; this leads to the implementation of DEMS on a field programmable gate array (FPGA). The FPGA is an integrated circuit designed to be configured by a customer or a designer after manufacturing – hence field programmable. The memory requirements of the SMG applications can be met by FPGA as the former involves computation with huge amount of data. It's a reprogrammable hardware development platform that harnesses the power of a dedicated high-capacity, low-cost programmable device to allow rapid interactive implementation and debugging of the designs.

A sample data string of RTDCU transmission is shown in Figure 5.28.

D_1 in Figure 5.28 is obtained from RTDCU 9 installed on Bus 3 of SMG shown in Figure 5.27; it represents the status of the battery – the status parameters of SoC and charging current. A detailed description of the various possibilities of data of D_1 is shown in Table 5.19. D_2 obtained from RTDCU 2 represents the current delivered to the PHS. If $I_{PH} \leq I_{PH(min)}$, then the status parameter, S_{W3} is assigned 0; else it is 1.

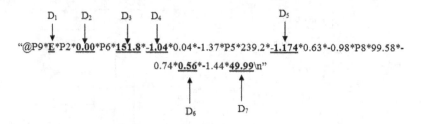

FIGURE 5.28 A sample real-time data string transmitted by the RTDCU.

TABLE 5.19

Description for D_1

D_1	Description	S_{SOC}	S_{lc}
A	Fully charged	1	0
B	To be charged	0	0
C	Fully discharged	−1	0
D	Charging	Previous value	1
E	Disconnected	Previous value	0
F	Discharging	Previous value	−1

D_3 and D_4 are extracted from RTDCU 6, installed on bus1, where the main grid is connected. D_3 represents the magnitude of the grid voltage (V_G) required to identify the mode of operation of SMG. D_4 represents the phase angle (δ_G) of V_G. **D_5** is extracted from RTDCU 5 installed on bus 5; it represents the phase angle (δ_{MG}) of the voltage on the microgrid side. These phase angles are used to identify the status of power exchange between the main grid and the microgrid.

Power exchange through the tie line (shown in Figure 5.29) is governed by Equation 5.17

$$P = \frac{V_G V_{MG}}{Z} \cos(\delta - \theta) - \frac{V_{MG}^2}{Z} \cos\theta \qquad (5.17)$$

where:

V_G: RMS voltage on Bus 1, V

V_{MG}: RMS voltage on Bus 5, V

δ: Power angle between the grid side and the microgrid side $= (\delta_G - \delta_{MG})$, °

$$Z = R + jX \; ; \; Z = \sqrt{R^2 + X^2}; \; \angle\theta = \tan^{-1}\frac{X}{R} \; ; \qquad (5.18)$$

Z: Impedance of the tie line connecting the grid and the microgrid, Ω,

X: Reactance of the tie line connecting the grid and the microgrid, Ω,

R: Resistance of the tie line connecting the grid and the microgrid, Ω, and

θ: Phase angle between the voltage and the current on the tie line, °.

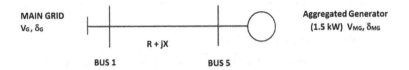

FIGURE 5.29   Tie line between the main grid and the microgrid.

The following operating conditions are identified with the tie line in Figure 5.29.

1. If $\delta_G > \delta_{MG}$, then microgrid is importing power from the main grid.
2. If $\delta_G < \delta_{MG}$, then microgrid is exporting power to the main grid.
3. If $V_G = 0$, then main grid is disconnected.

Figure 5.30 presents the algorithm to decide the status value for S_{W1}. V_{Gmin} is the minimum voltage required for the microgrid to operate in grid-connected mode.

D_6 extracted from RTDCU 8 represents the total load current (I_L) delivered to the aggregated load connected to the SMGS. If $I_L \leq I_{L(min)}$, then S_{W2} is assigned 0 and else 1. D_7 extracted from RTDCU 8 represents the frequency (f). This is the only RTDCU which transmits the frequency of the system along with the other data. Table 5.20 shows the summary of the data extracted from the respective RTDCUs and the associated status parameters.

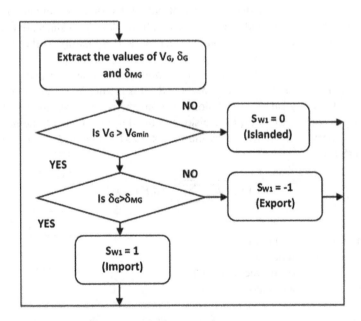

FIGURE 5.30   Algorithm for the status assignment of S_{W1}.

TABLE 5.20
Summary of Extracted Data

S. No.	Parameter	RTDCU No.	Status Parameter Associated with
1.	δ_G, V_G	RTDCU 6	S_{W1}
	δ_{MG}	RTDCU 5	
2.	I_L	RTDCU 8	S_{W2}
3.	I_{PH}	RTDCU 2	S_{W3}
4.	f	RTDCU 8	S_f
5.	I_c, SOC	RTDCU 9	S_{Ic}, S_{SOC}

5.6.1 DEMS IMPLEMENTATION ON FPGA

Given a training data set of the form (x_i, y_i), where $x_i \epsilon R^n$ is the ith example and $y_i \epsilon \{1 \ldots k\}$ is the ith class label, a learning model H is found such that $H(x_i) = y_i$ for new unseen examples. The problem is simply formulated in the two-classes case, where the labels y_i are just +1 or −1 for the two classes involved. The basic SVM supports only binary classification. The multiclass classification problem is decomposed into several binary classification tasks that are solved efficiently using binary classifiers.

Several methods have been proposed for such decomposition like one-against-all (OAA) and one-against-one (OAO). The OAA classification must train "k" binary SVM classifiers where k is the number of output decisions. The ith SVM classifier is trained with all samples of ith class as positive samples and takes all other examples to be negative samples. For the k number of output decisions, k numbers of binary SVM classifiers are generated. For a test data, all the decision values are computed by all decision functions and the output decision corresponding to the maximum value is the resulting decision.

In OAO classification, for every combination of two classes i and j, a corresponding binary SVM classifier is constructed. This results in the creation of $\frac{k(k-1)}{2}$ binary classifiers to yield $\frac{k(k-1)}{2}$ decision functions, where k is the number of output decisions. Since there are 7 DEM decisions in the developed model, the number of binary classifiers generated is 21. A multiclass SVM classifier $f(x)$ is defined by Equation (5.19) as

$$f(x) = b + \sum_{i=1}^{n} u_i d_i K(x, x_i) \tag{5.19}$$

where:
  b: Bias value of the classifier,
  u_i: Lagrange multiplier,
  d_i: Target output for the vector x_i, and
  $K(x, x_i)$: RBF kernel function.

$$K(x, x_i) = e^{-\gamma |x - x_i|^2} \tag{5.20}$$

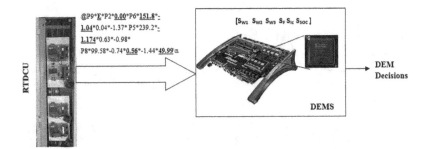

FIGURE 5.31   Data processing in DEMS.

The values of the Lagrange multipliers and bias are obtained during training of SVM. The OAO testing is performed by forming a direct acyclic graph (DAG).

Figure 5.31 summarizes the RTDCU data processing by DEMS. Once the decision is made, DEMS transmits the data to a server database in a pre-decided format as shown in Equation 5.21 where the characters "$" and "*" indicate the start and end of the data respectively; G/I indicates the mode of operation of SMG – G stands for grid-connected mode and I for islanded mode.

$$\$\delta_G, \delta_{MG}, V_G, I_L, I_{PH}, f, S_{W1}, S_{W2}, S_{W3}, S_f, S_{Ic}, S_{SOC}, G/I \text{ Decision} * \quad (5.21)$$

The received data in the data logger is time-stamped and stored in the database for future use. It is observed that the time interval between consecutive data logging is 4 s. This time interval envelopes:

- Receiving the data from the RTDCUs,
- Extracting the required data,
- Forming the status word,
- Classifying using SVM, and
- Transmitting the classified data to the server in the mentioned format.

5.6.2   DEMS ON GRID-CONNECTED SMG

The frequency excursions and the corresponding energy management actions taken by DEMS when SMG is in grid-connected mode are shown in Figures 5.32 and 5.33, respectively.

A small increase in grid frequency from 50 to 50.3 Hz as shown in Figure 5.32 is handled by increasing the speed of PHS and charging the battery together as shown in Figure 5.33. Similarly, a decrease in grid frequency from 50 to 49.99 Hz is handled by decreasing the speed of the PHS along with discharging the battery. The PHS is switched OFF eventually as the frequency is below the nominal value and is turned ON once the frequency reached the nominal value as seen in Figure 5.33.

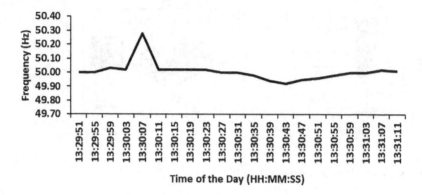

FIGURE 5.32   Frequency excursions in grid-connected mode.

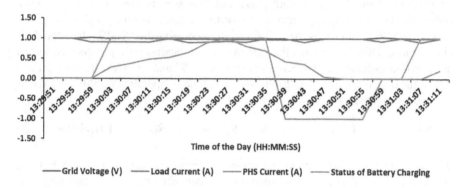

——Grid Voltage (V)   ——Load Current (A)   ——PHS Current (A)   ——Status of Battery Charging

FIGURE 5.33   DEM performed on SMG in grid-connected mode.

5.6.3   DEMS on Islanded SMG

The frequency excursions and the corresponding control actions in islanded mode are shown in Figures 5.34 and 5.35, respectively. It can be noticed when the frequency is very low that the battery discharges to the loads while the power delivered to the PHS is gradually reduced.

The LM performed as DR in this mode plays a significant role in maintaining the frequency. The decision of LM is obtained with an idle PHS and charging battery. The results of load management which involves inclusion and exclusion of local loads to/from the SMG is shown in Figure 5.36.

Once the consumers are disconnected from the SMG due to low frequency, the strategy for reconnecting to the SMG differs for solar-linked DR and non-DR consumers. The LM module can decide which DR consumer has to be included or excluded based on the frequency and the trend in frequency. Also, the non-DR consumers are connected to the SMG only after a stipulated amount of time (say after 30 min). This strategy motivates the non-DR consumers to become a member of solar-linked DR as the latter enjoys uninterrupted power supply with fiscal benefits.

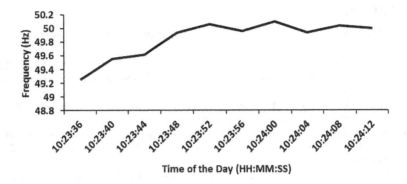

FIGURE 5.34 Frequency excursions in islanded mode.

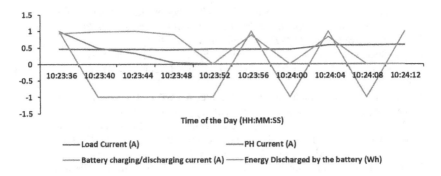

FIGURE 5.35 DEM performed on SMG in islanded mode.

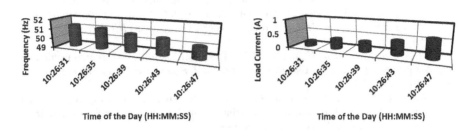

FIGURE 5.36 Results of load management.

When several such RE energized microgrids penetrate the conventional grid, such continuous actions taken on every SMG would render a significant assistance in maintaining the frequency within the permissible range.

The world will experience an extraordinary transformation in the incessant monotonous operation of the well-established legacy grid with the deployment of self-controlled regional microgrids in the near future; yet, the hindmost will continue its operation as it is vastly established to be deserted, thereby polishing the older technology with the latest advancements. It is therefore definite that an advanced power management technique like DEM is essential for the future power grid operation globally.

BIBLIOGRAPHY

1. Zaidi, A.A. and Kupzog, F. "Microgrid automation – A self-configuring approach." *Proceeding. INMIC 2008*, pp. 565–570.
2. Aghaei, J. and Alizadeh, M.-I. "Demand response in smart electricity grids equipped with renewable energy sources: A review." *Renewable and Sustainable Energy Reviews* 18 (2013): 64–72.
3. Ipakchi, A. and Albuye, F. "Grid of the future: Are we ready to transition to a smart grid?" *IEEE Power & Energy Magazine* 2009, pp. 52–62.
4. Chen, C., Duan, S., Cai, T., Liu, B. and Hu, G. "Smart energy management system for optimal microgrid economic operation." *IET Renewable Power Generation* 5(3) (2011): 258–267.
5. Hsu, C.-H., Chang, C.-C. and Lin, C.-J. "A practical guide to support vector classification." Department of Computer Science National Taiwan University, Taipei 106, Taiwan http://www.csie.ntu.edu.tw/~cjlin. Initial version: 2003 Last updated: April 15, 2010.
6. Kondolen, D., Ten-Hope, L., Surls, T. and Therkelsen, R.L. "Microgrid energy management system." *California Energy Commission Consortium for Electric Reliability Technology Solutions (CERTS) Consultant Report*, 2003.
7. http://techdocs.altium.com/display/HWARE/NanoBoard+3000+Series
8. www.hgpauction.com/auctiondata/233UniSolar.4/Technical_Data_Sheet_PVL_68.pdf
9. www.uie.org/sites/default/files/generated/files/pages/LoadManagement.pdf
10. www.cercind.gov.in/2015/draft_reg/Ancillary_Services.pdf
11. Manikandan, J., Venkataramani, B. and Avanthi, V. "FPGA implementation of support vector machine based isolated digit recognition system." *22nd IEEE International Conference on VLSI Design*, 2009, pp.347–352.
12. Nunna, H.S.V.S. and Doolla, S. "Energy management in microgrids using demand response and distributed storage—A multiagent approach." *IEEE Transactions on Power Delivery* 28(2) (2013): 939–947.
13. Luna, A.C., Diaz, N.L., Graells, M., Vasquez, J.C. and Guerrero, J.M. "Mixed-integer-linear-programming-based energy management system for hybrid PV-wind-battery microgrids: Modeling, design, and experimental verification." *IEEE Transactions on Power Electronics* 32(4) (2017): 2769–2783.
14. Mishra, S., Mallesham, G. and Jha, A.N. "Design of controller and communication for frequency regulation of a smart microgrid." *IET Renewable Power Generation* 6(4) (2012): 248–258.
15. Wang, X., Zhang, Y., Chen, T. and Giannakis, G.B. "Dynamic energy management for smart-grid-powered coordinated multipoint systems." *IEEE Journal on Selected Areas in Communications* 34(5) (2016): 1348–1359.
16. Zhang, Y., Jia, H.J. and Guo, L. "Energy management strategy of islanded microgrid based on power flow control." *IEEE PES Innovative Smart Grid Technologies (ISGT)*2012, pp. 1–8.

Index